ACOUSTIC FIELD ANALYSIS
IN SMALL MICROPHONE ARRAYS

Von der Fakultät für Elektrotechnik und Informationstechnik der
Rheinisch-Westfälischen Technischen Hochschule Aachen
zur Erlangung des akademischen Grades eines

DOKTORS DER INGENIEURWISSENSCHAFTEN

genehmigte Dissertation

vorgelegt von

Diplom-Ingenieur

Roman Scharrer

aus Meerbusch, Deutschland

Berichter:

Universitätsprofessor Dr. rer. nat. Michael Vorländer
Universitätsprofessor Dr.-Ing. Peter Vary

Tag der mündlichen Prüfung: 22. Juli 2013

Diese Dissertation ist auf den Internetseiten der Hochschulbibliothek online verfügbar.

Roman Scharrer

Acoustic Field Analysis
in Small Microphone Arrays

Logos Verlag Berlin GmbH

λογος

Aachener Beiträge zur Technischen Akustik

Editor:
Prof. Dr. rer. nat. Michael Vorländer
Institute of Technical Acoustics
RWTH Aachen University
52056 Aachen
www.akustik.rwth-aachen.de

Bibliographic information published by the Deutsche Nationalbibliothek

The Deutsche Nationalbibliothek lists this publication in the Deutsche Nationalbibliografie; detailed bibliographic data are available in the Internet at http://dnb.d-nb.de .

D 82 (Diss. RWTH Aachen University, 2013)

ISBN 978-3-8325-3453-0
ISSN 1866-3052
Vol. 14

Logos Verlag Berlin GmbH
Comeniushof, Gubener Str. 47,
D-10243 Berlin
Tel.:   +49 (0)30 / 42 85 10 90
Fax: +49 (0)30 / 42 85 10 92
http://www.logos-verlag.de

To

Clara and Linus

Abstract

In this work, the possibilities of an acoustic field analysis in small microphone arrays are investigated. With the increased use of mobile communication devices, such as smartphones and hearing aids, and the increase in the number of microphones in such devices, multi-channel signal processing has gained popularity. Apart from the definite signal processing, this thesis evaluates what information on the acoustic sound field and environment can be gained from the signal of such small microphone arrays.

For this purpose, an innovative sound field classification was developed that determines the energies of the single sound field components. The method is based on spatial coherences of two or more acoustical sensors and designed as extension of signal classification methods that are used for example in the automatic control stages of hearing aids. Two different approaches for the sound field energy estimation were researched. The method was successfully verified with a set of simulated and measured input signals. However, it showed to be very sensitive to sensor errors in this context, such as sensitivity mismatches of the sensors. To solve this problem, an adaptive automatic sensor mismatch compensation was developed, which proved able to fully compensate any slow sensor drift after an initial training.

Further, a new method for the blind estimation of the reverberation time based on the dependency of the coherence estimate on the evaluation parameters was developed. The method determines the reverberation time of a room from the spatial coherence between two or more acoustic sensors. Three different estimators, a neural network, as well as an empirical and an analytical model of the coherence estimate function, were proposed and evaluated to gain information on reverberation time, direct-to-reverberant energy ratio as well as signal-to-noise ratio. All methods show a good agreement of the estimated reverberation time with measured results. In rooms with non-exponential energy decay, the estimation results show the highest agreement with the measured early decay time.

Contents

Contents

Notation

x	signal in time domain
X	signal in frequency domain
$\boldsymbol{x}$	vector
$\boldsymbol{X}$	matrix
S_{xx}	power spectral density estimate for the signal $x(t)$
S_{xy}	cross spectral density estimate for the signals $x(t)$ and $y(t)$
$S(f) \cdot H(f)$	multiplication
$s(t) * h(t)$	convolution
$X(f)^*$	complex conjugate
$\mathcal{F}$	Fourier transformation
$|x|$	modulus of x
$\|\boldsymbol{x}\|$	Euclidian-norm of $\boldsymbol{x}$
$\langle\rangle$	average
$\hat{x}$	estimate of x

List of Symbols

E_{diffuse}	energy of signal component 'diffuse'
E_{free}	energy of signal component 'free'
E_{noise}	energy of signal component 'noise'
E_{reactive}	energy of signal component 'reactive'
E_{x}	energy of signal x
N	number of sources
P	source strength
S	room surface
T	reverberation time
V	room volume
Z	impedance
α_{ol}	overlap between two blocks in percent
α_c	decay rate for the exponential averge in spectral density estimations

List of Symbols

α	absorption coefficient
δ	Dirac impulse
$\hat{\gamma}_{xy}$	complex coherence estimate between the signals $x(t)$ and $y(t)$
γ_{pp}	spatial coherence between two pressure signals
γ_{xy}	complex coherence between the signals $x(t)$ and $y(t)$
erf	error function
e	exponential function
j	$= \sqrt{-1}$, the imaginary unit
ld	logarithms dualis
ω	$= 2\pi f$, angular frequency
$\overline{\alpha}$	mean absorption coefficient
ρ_0	density (of air)
III	Dirac comb function
θ	angle of sound incidence
$i(t)$	instantaneous intensity
i_c	complex intensity
i_r	reactive intensity
i_r	intensity in normal/radial direction
i	mean intensity
γ_{pu}	coherence between the sound pressure and the sound velocity at the same location
$\mathbf{e}_r$	unity vector in normal or radial direction
$\mathbf{u}_a$	part of the sound velocity in phase with sound pressure
$\mathbf{u}_d$	diffuse part of the sound velocity
$\mathbf{u}_r$	part of the sound velocity not in phase with sound pressure
$\mathbf{u}$	sound velocity or particle velocity
b	sensor sensitivity
c	speed of sound (in air)
d	distance between two receivers
f_s	sampling rate in samples per second
f_c	Schroeder frequency
f	frequency in Hertz
$g(t)$	abstract output signal
$h(t)$	impulse response (time domain) or transfer function (frequency domain)
k	The wave number $k = \omega/c = 2\pi/\lambda$

m	air absorption
$n(t)$	noise signal or spectrum
n_{bs}	block size in samples
n_{ol}	overlap in samples
p_a	direct part of the sound pressure
p_d	diffuse part of the sound pressure
p	sound pressure
r_c	critical distance
r	distance between source and receiver
$s(t)$	abstract input signal
t_{bs}	block size in seconds
t_c	time constant for the exponential averge in spectral density estimations
t	time in seconds
w_{diffuse}	energy density of the diffuse sound field
w_{free}	energy density of the direct sound
w_{reactive}	energy density of the reactive sound field
w	sound energy density
A	equivalent absorption area

Acronyms

AIR	Aachen impulse response database
BEM	boundary element method
BSFS	basic sound field sequence
BTE	behind the ear (hearing aid)
CEF	coherence estimate function
CTC	cross-talk cancellation
DRR	direct-to-reverberant energy ratio
EDT	early decay time
FEM	finite element method
FFT	(fast/discrete) Fourier transformation

HARTF	hearing aid related transfer function
HRTF	head related transfer function
ITC	in-the-canal (hearing aid)
PP	pressure-pressure
PSD	power spectral density
PU	pressure-velocity
SD	spectral density
SFC	sound field classification
SFI	sound field indicator
SNR	signal-to-noise energy ratio

1

Introduction

[The modern age] knows nothing about isolation and nothing about silence. In our quietest and loneliest hour the automatic ice-maker in the refrigerator will cluck and drop an ice cube, the automatic dishwasher will sigh through its changes, a plane will drone over, the nearest freeway will vibrate the air. Red and white lights will pass in the sky, lights will shine along highways and glance off windows. There is always a radio that can be turned to some all-night station, or a television set to turn artificial moonlight into the flickering images of the late show. We can put on a turntable whatever consolation we most respond to, Mozart or Copland or the Grateful Dead.

(STEGNER, 1971)

The modern world we live in is quite noisy. The amount of noise arising from mobility, with car traffic and airplane noise as well as trains, increases all over the world with no end in sight. Noise due to technical devices, as well as other peoples' 'noise', is almost everywhere, especially in cities, townships, factories and public places. Population growth leads to more and bigger cities. Mobility and wireless data transmission lead to the desideratum of communication everywhere and at any time. The rate of noise-induced diseases increases all over the world (BERGLUND and LINDVALL, 1995). But in a world of technical communication, noise is not only a problem for humans, communication devices also need to deal with noise (SLANEY and NAYLOR, 2011). There are people using their cellular phone in a crowded bar. In a scenario where you can barely understand the person next to you, telephony is still possible, thanks to modern signal processing methods. In every coffee place all around the world you see people talking to friends and colleagues by voice over IP or even video calls. Mobile communication in such situations seems to be desired by many people and technical devices need to cope with the problems that noise introduces. In the case of speech transmission, noise not only reduces speech intelligibility, but also decreases the performance of the coding used for the transmission channel. This leads to the situation where face-to-face communication may still be possible even though any communication using a phone is impossible. Even without the decrease of channel transmission performance, the speech intelligibility over a traditional communication channel is decreased (BENESTY, SONDHI, and HUANG, 2008). The signal from a mobile phone, or whatever device is used, is usually a single channel (mono) signal. Thus, speech intelligibility is even further reduced at the receiving end, bereaving the listener of the natural binaural hearing capability in humans that benefits them in distinguishing weak sounds in a noise and speech mixture (BLAUERT, 2005). Binaural transmission in such scenarios is often proposed and discussed, but despite its various advantages, it has not had its break through yet.

All the challenges faced by normal hearing persons are multiplied for people with a hearing loss (MCAULIFFE et al., 2012; VERMIGLIO et al., 2012). People who are able to cope with their hearing loss in normal situations may be completely helpless in situations with noise. They may not be able to follow a conversation or understand a word of their opponent. Technical aids like hearing aids are able to reduce the problem but are far from restoring the full functionality of human hearing. In fact, most hearing aid users are not completely satisfied with the performance of their devices in noisy situations (KOCHKIN, 2000). These people tend to avoid noisy situations. This also leads to their preclusion from attending public events like theatre or music performances or a night out with friends at a bar. People with a slight hearing loss tend to detach their hearing aids in noisy situations. This points out that there are still many problems to be faced by signal processing in hearing aids.

Despite the challenges, mobile devices are already capable of things nobody would have imagined a few years ago. The processor in a mobile phone is faster than that in a super computer two decades ago and that at an unbelievably small size and energy consumption. Thanks to that computational power, a wide range of signal processing strategies is available in mobile phones, hearing aids and other communication devices to reduce noise, increase speech intelligibility and finally increase user satisfaction. Most of these strategies are not beneficial in all situations (NAYLOR and GAUBITCH, 2012). They tend to disturb the signal, introduce artifacts or simply sound unnatural. So, if these strategies are not adjusted correctly, they might even disturb the user. The performance of most signal processing strategies depends to a large extent on several boundary conditions like the type of signal, the sound field and the user's preferences and expectations. Accordingly, the correct determination of such conditions is crucial to the total performance of any communication device.

Recently, the advantages of multi-sensor input for noise suppression and speech enhancement have led to communication devices with multiple acoustic sensors (HAMACHER et al., 2005). High performance hearing aids utilize two or more microphones to employ beamforming techniques to focus on the counterpart speaker (HEESE et al., 2012). Mobile phones come with two or even three microphones that are mainly used for background noise suppression. Some notebooks also carry multiple sensors for such purposes. With the advent of such small microphone arrays, the question becomes obvious what further advantages can be gained from the additional input signals. A small microphone array in this case is the combination of two or more microphones that are placed close by in one device, in the range of a few millimeters to a few centimeters. This is a contrast to classic microphone arrays used in acoustic measurements with tens to hundreds of sensors and array dimensions of some meters.

Figure 1.1 shows the sound field a small microphone array may be exposed to in a room. There is a speaker but also a concurring source in form of a loudspeaker. While the speaker usually emits a speech signal, a loudspeaker can reproduce virtually any form of signal, from speech over music to noise and technical sounds and also any combination of such. The signals from both sources travel on a direct path from the source to the microphone array. But there are also reflections from the floor and from the walls. Further, a diffuse sound field without any inbound direction is present, partly from the rooms' reverberation, partly from other noise sources. Depending on the frequency, the sound transmission is best described by a wave propagation or a ray propagation. Finally, the sensors of the microphone array are not perfect, so they also add some noise to the signal.

The signals received by the microphone array are a combination of some factors: the number, strength and position of the sound sources, the transfer paths from the sources to the receivers with some early reflections, as well as late, possibly diffuse, reverberation. There may also be additional acoustic noise from other sources. Finally, there is some noise emerging at the sensors. Thermic noise is always present in acoustic sensors and amplification devices. There may also be noise due to wind turbulence near the microphones, or mechanical noise due to contacts of the housing with cloth or other objects.

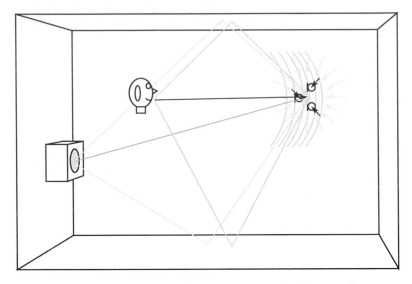

Figure 1.1: Sound field in a room with two sources perceived by a small sensor array

1.1 Objectives

The main objective of this thesis is to describe methods that are able to gather information on the acoustic environment based on the input signals of a small microphone array. The sound field that is perceived by the receiver is composed of different field types. A sound field classification not only determines the dominant type of sound field, but also estimates the energy distribution between the different sound field types. Further, a method for the blind estimation of the reverberation time, the signal-to-noise energy ratio (SNR) and direct-to-reverberant energy ratio (DRR) is proposed.

The main question about the acoustic environment can be separated in two categories:

- The local component: What sound field are the sensors exposed to? How is the sound field around the sensors composed? And how is it perceived?

- The global component: What are the (acoustic) room properties? What is the reverberation time?

This thesis will try to gather some of that information from the input signal of a small sensor array without any a priori knowledge. The local component is analyzed by a sound field classification that gives information on the local composition of the sound field in the area of the microphone array. Further room acoustic properties are gather by a blind reverberation time estimation that also delivers the SNR and DRR of the current scene.

Objectives that are not addressed in this thesis are the estimation of the number and position of sound sources, a classification of the sources' signals or the evaluation of user preferences, like which sound source he /she wants to listen to.

1.2 Overview of the Thesis

Some of the basics that should help the reader comprehend this thesis are presented in Chapter 2. The chapter explains some basics on room acoustics, the basic sound fields and how typical room acoustic measurements are carried out. It also introduces the hearing aid dummy head and the sound field microphone used in some evaluations and experiments within this thesis. Apart from measurements, some methods for the simulation of acoustic fields and room acoustics are introduced. Some basic concepts on spatial coherence are discussed, as spatial coherence is the primary indicator for the sound field classification, as well as the reverberation time estimation introduced in this thesis. A new function, called the coherence estimate function (CEF), is defined. Finally, some insights on the sound field indicators are given, also with respect to their computation in real time.

Chapter 3 introduces the sound field classification based on the sound field indicators. After a brief problem statement and the definition of the necessary boundary conditions like the number of acoustic sensors, two different approaches for a classification are introduced and verified. Error influences like sensor mismatch and their automatic compensation are discussed. The section closes with some examples of sound field classifications in different acoustic situations.

A new method for the blind estimation of the reverberation time is introduced in Chapter 4. The method is based on the influence of the reverberation time and other parameters on the CEF. The influences are studied in detail in Section 4.1. Three different approaches for the estimation of the reverberation time from the CEF are introduced. Some important aspects on the implementation in real time scenarios are discussed and some error sources and a possible post-processing of the results are introduced. The chapter closes with a verification of the three methods with measured binaural impulse responses.

Finally, Chapter 5 summarizes the results of the thesis, points out the contributions of the author and presents an outlook for possible further research on the discussed topics.

2

Theoretical Essentials

The following chapter will explain the theoretical essentials necessary for the acoustic field analysis described in this thesis. The essentials include some basics on room acoustics, the measurement of acoustic systems and possibilities of the simulation of such.

2.1 Signal Processing in Mobile Devices

Many mobile devices are used within a communication context. Mobile phones are designed for telephony (although they may be used extensively for other purposes). Hearing aids aim to reduce the limitations in speech intelligibility of hearing impaired persons. All those devices receive, process and transmit audio signals such as speech. The mobility of these devices also means that they can used in the most different places. The acoustic situation in many of those places is often not optimal for speech intelligibility. There may be other sound sources, concurring speakers and reverberation. To allow any communication in such situations using a mobile phone or a hearing aid, the signal has to be processed to increase speech intelligibility. The methods utilized for this purpose will be explained on the example of modern hearing aids. The methods used in mobile phones and other devices are very similar.

Most modern hearing aids use multiple microphones as input sensors. Depending on the size of the hearing aids, two or three microphones are typical instrumentations. Figure 2.1 shows the typical stages of signal processing in modern hearing aids. The calculation is separated into two blocks: the actual signal processing and a classification system that controls the signal processing. On the side of the signal processing, the multi-microphone input is used for beamforming to reduce noise from surrounding sources and to focus on the opposite speaker. Feedback suppression is necessary to keep the system stable despite the very high gain and output level. The subsequent processing is usually done in frequency

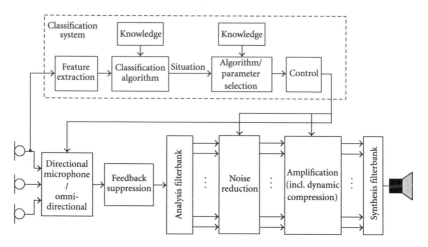

Figure 2.1: Processing stages of a high-end hearing aid (from (HAMACHER et al., 2005))

bands, which are generated by an analysis filter bank. After the processing, these frequency bands are assembled by a synthesis filter bank. Among the typical signal processing methods are noise reduction and amplification together with a dynamic compression.

The single algorithms and the range of their possible settings work in a situation-specific way (HAMACHER et al., 2005). A noise reduction, for example, may have a negating influence on the sound of music, although it is obviously beneficial in a situation with a lot of noise. Accordingly, the different stages of the signal processing need to be controlled, switched and adjusted according to situation. For this purpose, a classification and control segment adjusts the signal processing to the situation. The aim of the classification stage is to gather information on the signal, the situation, the acoustic environment and the user's preferences, and accordingly adjust the signal processing so that the best possible performance and user satisfaction is achieved. As the room acoustic has a significant influence the corresponding parameters and theories will be described in more detail in the following section.

2.2 Room Acoustics

KUTTRUFF (2000) gives a very broad introduction into general room acoustics. Further information on (room) acoustic sound fields in the context of sound intensity measurements can be found in (FAHY, 1989).

A multitude of parameters is used to describe the acoustic properties of a room and the sound field perceived by a receiver. The reverberation time T is a prominent example. It describes the time in which the energy density in a room drops 60 dB after a sound source is switched off. The reverberation time directly influences the speech intelligibility in a room as well as the performance of many signal processing strategies (BLAUERT, 2005).

An obvious influence on the sound field is given by the room geometry, as it directly influences the reflections from walls and other objects. The room volume V and surface S are determined by the room geometry. The number N, position and level of the present sound sources relative to the receivers also change the perceived sound field. The more (uncorrelated) sound sources are active, the more diffuse the sound field will be. Further, the distance r between source and receiver influences the amounts of direct and diffuse energies at the receiver location.

Using statistical considerations, the reverberation time can be calculated from the room volume V, the room surface S and the average absorption coefficient $\overline{\alpha}$ as (EYRING, 1930)

$$T = \left(-0.163\frac{\text{s}}{\text{m}}\right) \cdot \frac{V}{S \cdot \ln\left(1 - \overline{\alpha}\right)} \tag{2.1}$$

The absorption coefficient of most materials is frequency dependent, as is the air absorption m, which is not considered in (2.1). Accordingly, the reverberation time of a room is typically frequency dependent and is usually quoted in octave or third octave bands (DIN-EN-ISO-266, 1997; ISO-3382-1, 2009).

Depending on T and V, the sound field in a room is composed of discrete modes at lower frequencies. For higher frequencies, modes exist as well, but at each frequency many modes overlap in a rather chaotic way, so that the resulting sound field can be assumed diffuse. The *Schroeder* or critical frequency f_c is the typical crossover frequency between modal behavior and (quasi) diffuse sound fields. It can be calculated as (SCHROEDER, 1996)

$$f_c = 2000\sqrt{\frac{T}{V}} \tag{2.2}$$

The sound energy density in any sound field is defined as (FAHY, 1989)

$$w = \frac{1}{2}\rho_0||\boldsymbol{u}||^2 + \frac{p^2}{2\rho_0 c^2} \tag{2.3}$$

, where p is the sound pressure and $\boldsymbol{u}$ is the sound velocity.

The propagation of sound energy is called intensity. The instantaneous sound intensity $\boldsymbol{i}(t)$ is defined as

$$\boldsymbol{i}(t) = p\boldsymbol{u} \tag{2.4}$$

The mean intensity $\boldsymbol{i}$ over a time period T can be calculated as

$$\boldsymbol{i} = \frac{1}{T}\int_0^T p(t)\boldsymbol{u}(t)\mathrm{d}t \tag{2.5}$$

$$= \frac{1}{2}\Re\{p\boldsymbol{u}^*\} \tag{2.6}$$

$$= \frac{1}{2}\Re\{\boldsymbol{s}_{pu}\} \tag{2.7}$$

As the two agents of energy flow, pressure and particle velocity, are not necessarily in phase, there may also be some reactive component in the energy flow, just as in electrical circuits, where a phase shift between voltage and current may lead to a reactive power component.

Accordingly, the complex intensity $\boldsymbol{i}_c$ can be written as a summation of the mean intensity $\boldsymbol{i}$ and the reactive intensity $\boldsymbol{i}_r$

$$\boldsymbol{i}_c = \boldsymbol{i} + \mathrm{j}\boldsymbol{i}_r \tag{2.8}$$

$$= \frac{1}{2}p\boldsymbol{u}^* \tag{2.9}$$

$$= \frac{1}{2}\boldsymbol{s}_{pu} \tag{2.10}$$

2.2.1 Basic Sound Fields

The basic sound fields describe idealized sound fields that are completely defined by basic attributes and are often used as assumption for analytical calculations and as boundary conditions for signal processing. The basic sound fields in a pure form are seldom observed in realistic environments, but are often a good approximation for parts of the sound field (FAHY, 1989; JACOBSEN, 1989). Accordingly, many sound fields, as a simplification, can be expressed as a superposition of the basic sound fields.

The *free* sound field describes a free and undisturbed wave propagation. This also applies to conditions outside rooms, or inside rooms with sparse reverberation. It is a valid approximation for the direct sound from a source in any room.

Accordingly, the sound pressure p and the sound velocity $\boldsymbol{u}$ for one mono frequent point source in free field conditions can be written as (FAHY, 1989)

$$p(r,t) = \frac{B}{r} e^{j(\omega t - kr)} \tag{2.11}$$

$$\boldsymbol{u}(r,t) = \frac{b}{\omega \rho_0 r} \left(k - j/r \right) e^{j(\omega t - kr)} \boldsymbol{e}_r \tag{2.12}$$

$$= \frac{p(r,t)}{\rho_0 c} \cdot \left(k - j/r \right) \boldsymbol{e}_r \tag{2.13}$$

where k is the wave number and B can be calculated from the source strength Γ as

$$B = j \sqrt{\frac{\rho_0 c}{2\pi}} P. \tag{2.14}$$

$\boldsymbol{u}$ can be separated into one part $\boldsymbol{u}_a$, which oscillates in phase with p (and accordingly only contributes to i), and one $\boldsymbol{u}_r$, which oscillates with a phase shift of $\pi/2$ (and only contributes to i_r)

$$\boldsymbol{u}(r,t) = \frac{p(r,t)}{\rho_0 c} \cdot k \boldsymbol{e}_r + \frac{p(r,t)}{\rho_0 c} \cdot (-j/r) \boldsymbol{e}_r \tag{2.15}$$

$$= \boldsymbol{u}_a(r,t) + \boldsymbol{u}_r(r,t). \tag{2.16}$$

Using (2.11) and (2.12) in (2.3) leads to a sound energy density of (BERANEK,

1993)

$$w = \frac{B}{\rho_0 c^2 \cdot r} \left(1 + \frac{1}{2k^2 r^2}\right). \tag{2.17}$$

By using (2.15) instead, the sound energy density can be separated into an active (w_{free}) and a reactive part (w_{reactive})

$$w = \frac{1}{2}\rho_0 |u_a|^2 + \frac{p^2}{2\rho_0 c} + \frac{1}{2}\rho_0 |u_r|^2 \tag{2.18}$$

$$= \frac{B}{\rho_0 c^2 \cdot r} + \frac{B}{\rho_0 c^2 \cdot r} \frac{1}{2k^2 r^2} \tag{2.19}$$

$$= w_{\text{free}} + w_{\text{reactive}} \tag{2.20}$$

For distances $r \gg k$, the reactive term can be neglected.

The sound intensity i_c in a sound field radiated from a point source can be written as

$$i = \frac{B^2}{2r^2 \rho_0 c} \left(1 + \cos 2\left(\omega t - kr\right)\right) \mathbf{e}_r \tag{2.21}$$

$$i_r = \frac{B^2}{2r^3 \rho_0 \omega} \sin 2\left(\omega t - kr\right) \mathbf{e}_r \tag{2.22}$$

$$i_c = i + \mathrm{j} i_r. \tag{2.23}$$

Apart from the strong reactive effect in the near field of a sound source, *reactive* sound fields also occur in standing waves and modes.

The model of a *diffuse* sound field is often used to describe the sound field in reverberant rooms or a sound field generated by multiple uncorrelated sound sources.

A diffuse sound field can be described as the superposition of infinitely many point sources uniformly distributed on a sphere emitting uncorrelated signals.

Accordingly, p and $\boldsymbol{u}$ can be written as

$$p(r,t) = \sum_{n=1}^{\infty} \frac{B_n}{r_n} e^{j(\omega t - k r_n)} \tag{2.24}$$

$$\boldsymbol{u}_r(r,t) = \sum_{n=1}^{\infty} \frac{B_n}{\omega \rho_0 r_n} \left(k - j/r_n\right) e^{j(\omega t - k r_n)} \mathbf{e}_{r,n}. \tag{2.25}$$

The energy density of the diffuse sound field w_{diffuse} only depends on the equivalent absorption area $A = S\overline{\alpha}$ and on the source strength P, and can be written as (KUTTRUFF, 2000)

$$w_{\text{diffuse}} = \frac{4P}{c \cdot A}. \tag{2.26}$$

In a diffuse sound field no energy propagation takes place, as the energy distribution is uniform in the whole space. This means that although the instantaneous intensity $\boldsymbol{i}(t)$ may behave randomly, the long term intensity $\boldsymbol{i}$ is zero.

The ratio between w_{free} and w_{diffuse} is called DRR and can be written as

$$\text{DRR} = \frac{w_{\text{free}}}{w_{\text{diffuse}}} \tag{2.27}$$

$$= \frac{A}{16\pi r^2} \tag{2.28}$$

$$= \frac{0.163V}{16\pi r^2 \cdot T}. \tag{2.29}$$

That means, in a room with only one source, the DRR is directly proportional to $1/r^2$.

The distance where the sound energy density of the direct sound is equal to that of the diffuse sound field, so that $w_{\text{free}} = w_{\text{diffuse}}$, is called critical distance r_c with (SCHROEDER, 1996)

$$r_c = \sqrt{\frac{S \cdot \alpha}{16\pi}} \tag{2.30}$$

$$\approx 0.057 \cdot \sqrt{\frac{V}{T} \cdot \frac{\text{s}}{\text{m}}} \tag{2.31}$$

substituting $\frac{S \cdot \alpha}{16\pi}$ with $r_c{}^2$ in (2.28) leads to

$$\mathrm{DRR} = \left(\frac{r_c}{r}\right)^2. \tag{2.32}$$

In the strict sense, those definitions are only valid for stationary signals with one dominant sound source.

The basic sound fields, *free*, *reactive* and *diffuse*, are sound fields, whereas *noise* does not describe a real sound field but rather the absence of any acoustic sources. For the case of a non-ideal receiver like a microphone, there will always be additional electronic noise from the sensor itself, the preamps and the analog-to-digital converter. In a strict sense, this is no sound field but a sensor artifact. For the subsequent signal processing it cannot be distinguished from a pressure signal. The ratio between the energy E_{signal} of any acoustic signal and the energy of the noise E_{noise} is called SNR and can be calculated as

$$\mathrm{SNR} = \frac{E_{\mathrm{signal}}}{E_{\mathrm{noise}}}. \tag{2.33}$$

There are various definitions of the SNR, mostly differing in the point what is considered signal and what is considered noise. In this thesis, every acoustic signal will be considered wanted, whereas only noise emerging at[1] or in[2] the receiver will be considered noise.

2.2.2 Combined Sound Fields

In a room, the sound pressure from one point source and the reflections from the room surfaces can be written as

$$p(r,t) = \frac{B_1}{r_1} \mathrm{e}^{\mathrm{j}(\omega t - kr_1)} + \sum_{n=2}^{\infty} \frac{B_n}{r_n} \mathrm{e}^{\mathrm{j}(\omega t - kr_n)} \tag{2.34}$$

$$\begin{aligned} \boldsymbol{u}(r,t) =& \frac{B_1}{\omega \rho_0 r_1} \left(k - \mathrm{j}/r_1\right) \mathrm{e}^{\mathrm{j}(\omega t - kr_1)} \boldsymbol{e}_r \\ &+ \sum_{n=2}^{\infty} \frac{B_n}{\omega \rho_0 r_n} \left(k - \mathrm{j}/r_n\right) \mathrm{e}^{\mathrm{j}(\omega t - kr_n)} \boldsymbol{e}_{r,n} \end{aligned} \tag{2.35}$$

---

[1] wind noise or mechanical noise like scratching
[2] electronic noise

An acoustic sensor with a sensitivity b will finally sense either sound pressure or sound velocity. As the sensor is never ideal, it will always add some electronic noise to the signal, as will the amplifier and the AD-converter. The final signal for a pressure sensor can then be written as

$$s_p(r,t) = b \cdot p(r,t) + n_p(t) \qquad (2.36)$$

or for a velocity sensor

$$s_u(r,t) = b \cdot \boldsymbol{u}(r,t) + n_u(t). \qquad (2.37)$$

Most velocity sensors only sense the sound velocity in one direction $\mathbf{e}_n$. Accordingly, the signal perceived can be written as

$$s_u(r,t) = b \cdot \boldsymbol{u}(r,t) \cdot \mathbf{e}_n + n_{u,r}(t) \qquad (2.38)$$

2.2.3 Room Acoustic Measurements

For the verification of many signal processing strategies, examples are very helpful. Recordings of different situations as well as room impulse responses that can be convolved with different signals are prominent examples of tools necessary for the evaluation of signal processing strategies. The DIN-EN-ISO 3382 (2006) includes a general guidance on how room acoustic measurements should be carried out.

Transfer function measurements

Measuring transfer functions is one of the most important daily tasks in all areas of acoustics. Especially in room acoustics, when convolving impulse responses with anechoic audio material, a very high dynamic range is necessary (DIETRICH and LIEVENS, 2009). The most reliable method to get a very high dynamic range measurement in a reasonable time is a transfer-function measurement using a sweep excitation signal (FARINA, 2000)(MÜLLER and MASSARANI, 2001). Using a sweep excitation has the main advantage that most non-linear distortions, for example from the non-ideal loudspeakers, are visible in the impulse response and can be separated and eliminated. Further, the sweep excitation methods is far more robust to time variances than, for example, maximum-length-sequences (MLS) (MÜLLER and MASSARANI, 2001). The use of a sweep excitation signal

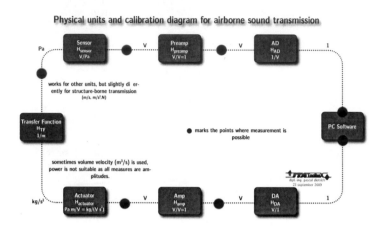

Figure 2.2: Calibration of the measurement chain (FONSECA et al., 2010)

ensures a maximal utilization of power amplifier and speaker, and generates the best SNR possible with the given setup and time (MÜLLER and MASSARANI, 2001).

For all measurements in this context, *Matlab* is used as measurement system, using the *ITA-Toolbox* (DIETRICH, GUSKI, et al., 2013), an acoustic measurement and signal processing function set. The *ITA-Toolbox* is able to do absolute measurements, using the calibration method shown in Figure 2.2. Using the known combined amplification factors and sensitivities of the hearing aid microphones in the dummy head and a known power amplifier and speaker, absolute room impulse responses can be measured and absolute recordings can be done, giving a high precision for later testing of signal processing.

Reverberation time estimation from room impulse responses

The process of a reverberation time calculation (or rather estimation) from a room impulse response is described in DIN-EN-ISO 3382 (2006). The corresponding methods are implemented in the *ITA-Toolbox* and all non-blind calculations of the reverberation time were carried out using this state-of-the-art implementation.

Hearing aid dummy head

For hearing aid specific measurements and recordings, a dummy head equipped with hearing aids was built as indicated in Figure 2.3. The head and ears are identical to the ITA artificial head. Instead of microphones in the ear channel, the dummy head is equipped with four hearing aids, one behind the ear (hearing aid) (BTE) and one in-the-canal (hearing aid) (ITC) hearing aid shell at each ear. The hearing aid shells include their original microphones, which are connected to a digital audio device. The hearing aid dummy head can be used for measurements, just like a usual dummy head. Instead of a binaural signal, the result will be an eight-channel recording with the signals of four hearing aids, each containing two microphones. This signal can be used for online or offline performance tests of hearing aid signal processing as well as demonstrations of hearing aid behavior.

The BTE hearing aids include two microphones with a distance of 12.4 mm. The ITC hearing aids also include two microphones but with a closer distance of 6.7 mm.

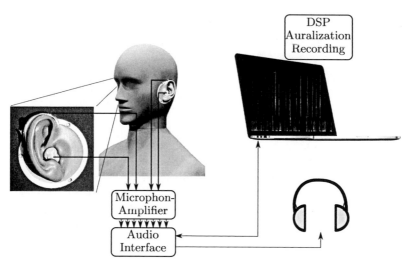

Figure 2.3: Sceme and photo of the hearing aid dummy head

The hearing aid microphones can be calibrated using a broadband calibrator. The calibrator is shown in Figure 2.4. It uses a headphone driver and an electret microphone (Sennheiser KE-4) as reference. It is designed as a pressure chamber. Its dimensions are chosen in a way to ensure a reliable calibration for frequencies up to 10 kHz. The calibrator uses soft foam to ensure an air tight volume when pressed to a hearing aid.

One problem with this kind of calibration is the connection between the calibrator and the microphone. Due to the placement of the microphones inside the hearing aids, it is very complicated to get an airtight connection between calibrator and microphone. Further tests are necessary to verify the calibration process using this calibrator.

Other possible methods are free field calibration and diffuse field calibration, using a reference microphone (KUTTRUFF, 2007). The main problem with both methods is that the microphones are fixed in the hearing aid housing, so the influence of the housing will always appear in the calibration process.

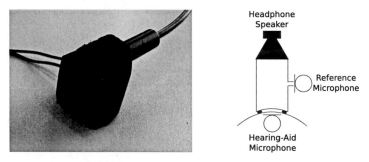

Figure 2.4: A broadband calibrator for hearing aid microphones

The channel sensitivities of the firewire audio interface used in the hearing aid dummy head have been measured using a voltage source that emits a sinus wave at 1000 Hz and 1 V. The audio interface shows channel sensitivities according to Table 2.1. Channels 1 and 2 of the audio interface are microphone inputs and show a slightly higher sensitivity than the line level inputs 3-8. The channel sensitivities of the hearing aid preamps used in the hearing aid dummy head have been measured by measuring the transfer function from input to output. Channels 1 and 2 of the audio interface are microphone inputs and have higher a impedance than the line inputs 3-8. This seems to affect the amplification of the preamps, as channels 1 and 2 have a lower amplification. The sensitivities are shown in Table 2.1, the frequency responses in Figure 2.5.

Channel Number	Preamp Sensitivity in V/V	AD Sensitivity in 1/V
1	7.81	0.285
2	7.80	0.285
3	9.54	0.199
4	9.52	0.199
5	9.28	0.201
6	9.28	0.201
7	9.33	0.201
8	9.30	0.201

Table 2.1: Channel sensitivities of preamps and AD

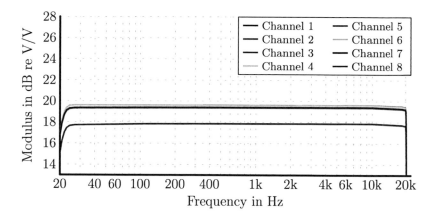

Figure 2.5: Frequency responses of the hearing aid preamplifiers

The compensation for absolute, calibrated measurements with the eight hearing aid microphones is shown in Figure 2.6.

For this setup, a hearing aid related transfer function (HARTF) has been measured in a spectral resolution of $1 \times 5\,°$. The measurement has been carried out according to the head related transfer function (HRTF) measurement method described in (LENTZ et al., 2007). This HARTF can be used as replacement for the typical HRTF in room acoustic simulation models and auralization with a hearing aid context.

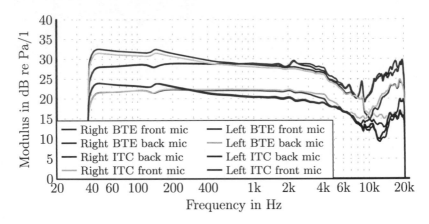

Figure 2.6: Final compensation to compensate for the non perfect frequency response of the hearing aid microphone sensitivities along with the amplification preamps. This compensation allows absolute measurements with the hearing aid dummy head.

Sound field microphone

Apart from the hands-on example of the hearing aids, a more general small microphone array was used for some evaluations. A sound field microphone, consisting of four omnidirectional pressure sensors placed in a sphere with a diameter of 14 mm. The microphones form the shape of a regular tetrahedron. A picture of the sound field microphone is displayed in Figure 2.7.

Figure 2.7: Sound field microphone

2.3 Room Acoustic Modeling and Simulation

For the development and verification of a sound field classification and analysis, a set of sound field models, as well as examples for room acoustic situations, are necessary. The basic sound fields are seldom experienced in a pure form. Measurements or recordings of pure basic sound fields are practically impossible. Models of the basic sound fields were created for the testing of the sound field classification. The implementation of these models is described in following section. The basic sound field sequence (BSFS), which is used for the evaluation of the sound field classification, is also described.

For the simulation of real sound fields, a variety of simulation methods is available. The advantages and disadvantages of the different methods are discussed with a focus on applicability in the context of small microphone arrays. Finally, the stochastic room impulse generation method is explained, which was used for Monte-Carlo simulations.

2.3.1 Basic Sound Field Models

The following section describes how the basic sound fields introduced in Section 2.2.1 can be modeled for validation of sound field classification methods.

Free Free field conditions can be simulated by an implementation of (2.11) and (2.12). It is important to choose the distance r big enough, so that no significant reactive components occur[3]. Sometimes it is necessary to include a receiver directivity. This can be achieved by a convolution of the receiver's directivity with the signal at the receiver (without directivity). A directivity is usually employed to include the influence of the sensor housing or the human head (this directivity is called HRTF).

Diffuse A perfectly diffuse sound field would consist of infinitely many uncorrelated sound waves coming equally distributed from all directions (JACOBSEN and ROISIN, 2000) as described in (2.24) and (2.25). As it is not possible to simulate this infinite number of waves, a large but finite number of uncorrelated waves from random directions is a valid approximation for a diffuse sound field.

---

[3] Of course, the model can also neglect the reactive component, independent from the distance

Reactive A completely reactive sound field can be simulated by a 90° shift between sound pressure and sound velocity. Utilizing (2.11) and (2.12) neglecting the 'in-phase' part of u leads to a purely reactive sound field.

Noise Noise can be simulated as uncorrelated random signals for the different sensors. Noise is usually normal distributed so the random signal should have the same distribution. The noise signal is not influenced by the sensor housing or head in any way.

The basic sound field sequence

The BSFS is a convenient method for the validation of the different stages of the sound field classification. It is a sequence of the basic sound fields, one after the other. It can also be used to investigate different influences like microphone mismatch on the classification. The main advantage is that in each segment the classification should return a perfectly defined value. Differences from that perfect behavior can easily be quantified.

The sequence in which the basic sound fields occur is:

Figure 2.8: The BSFS

2.3.2 Room Acoustic Simulation

Whenever a room acoustic measurement is not feasible or advisable, maybe due to time or complexity reasons, or because the room does not even exist, a room acoustic simulation is a possible solution for auralization or evaluation. To determine the transfer function, or impulse response, from an acoustic source to a receiver in an enclosed space, a room acoustic model is necessary. Figure 2.9 shows a primary classification of room acoustic models. There are vast differences between the different methods, in calculation time, (audible) accuracy and physical exactness. The applicability of some methods also depends on the room and the frequency range of interest. In general, wave based approaches are

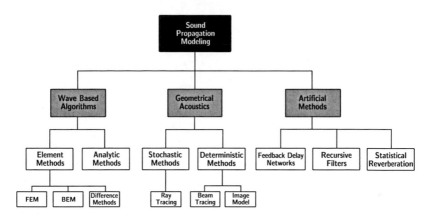

Figure 2.9: Classification of room acoustic models (VORLÄNDER, 2008)

better suited for low frequency applications, with $f \leq f_c = 2000\sqrt{\frac{T}{V}}$. For higher frequencies with $f > f_c$, geometrical methods are better suited. There are also approaches of a combination of both methods to get realistic broadband impulse responses (ARETZ, 2012; ARETZ et al., 2009).

A very broad overview and discussion on the different models can be found in (VORLÄNDER, 2008).

Wave based algorithms

Wave based algorithms include the finite element method (FEM), boundary element method (BEM), difference methods (e.g. finite time difference method (FTDM)), and analytic methods.

For very simple room geometries, an analytic solution of the Helmholz equation is possible. For typical rooms this approach is not suitable due to too complex geometries (KUTTRUFF, 2000).

The finite element method works by a discretization of a field volume into volume elements. In these elements, the energy formulation of the harmonic field is used. The principle is very general and is applied in many fields, like mechanical problems, fluid dynamics, heat conduction, electromagnetic or acoustic field problems (VORLÄNDER, 2008). For an FEM simulation, a room model with geometric information as well as well-defined boundary conditions is necessary. The quality of the result crucially depends on the quality of the input data (ARETZ, 2009; ARETZ, 2012). The finite element method is very well suited for frequencies below f_c. For higher frequencies, the results tend to differ significantly from the reality. Although the method is theoretically also correct for high frequencies, the uncertainties in the model and boundary conditions lead to high errors compared to measurements. In addition, the computational load increases significantly with the frequency. The finite element method can be very time consuming. In most cases, it is not suitable for real-time applications or Monte-Carlo simulations.

The boundary element method is mostly used in radiation and scattering problems that are characterized by boundary conditions like local impedances, admittances or surface velocities. The boundary element method is usually not used in the simulation of room acoustics but rather for the simulation of directivities like the HRTF.

Geometrical acoustics

Geometrical acoustics include methods like ray tracing, which is also very popular in computer graphics, beam tracing and mirror source models. All those methods show good results for frequencies above f_c. They are not suitable for a realistic representation of modal behavior. The mirror source method is very accurate for the first reflections, but it cannot model diffraction. Additionally, the number of mirror sources grows exponentially as a function of reflection order, resulting in a very high computation time for long impulse responses. Faster methods for the diffuse tail of the reverberation are ray or beam tracing methods. Often both methods are combined, using mirror sources for accurate early reflections and ray tracing for the reverberation tail. One example is the real-time capable room acoustic auralization system *RAVEN*, which was developed by SCHRÖDER (2011).

Problems specific to the simulation of microphone arrays

For all wave based algorithms, a model of the microphone array is necessary. In addition, the model must be fine enough to represent the small microphone distances. Apart from that, all wave based methods are suitable for a realistic simulation of the input signals of a high precision microphone array, including all effects like source near field effects and diffraction at the receivers.

For geometrical acoustic approaches, usually an HRTF or a directivity database is used. This HRTF can be measured, for example, for a hearing aid setup on a dummy head and therefore theoretically be incorporated into the geometrical models. For other setups of microphone arrays a directivity can be measured or simulated and used for the simulation just like an HRTF. One remaining problem is, that most commercial as well as non-profit geometrical acoustics models only accept HRTFs with two channels, although four would be necessary for a binaural hearing aid setup or a sound field microphone. Repeated simulations with a split HRTF, on the other hand, show wrong results of spatial coherences as well as other sound field indicators, due to stochastic processes in the ray tracing methods. Further, most methods are only capable of simulating sound pressures, so the sound velocity can only be estimated from a sound pressure distribution at some points around the point of interest.

Binaural stochastic room impulse response

The following method was already introduced in (SCHARRER and VORLÄNDER, 2010). The respective part is repeated here for the reader's convenience.

Most room acoustic simulation methods need a room model to work with. That means the room geometry as well as the boundary conditions have to be known. Another possibility is the stochastic simulation of a room. In this case, no room model is necessary. The necessary information on the room are the room volume V, the room surface S, the source position, relative to the receiver and the average absorption coefficient α, the reflection factor R or the reverberation time of the room T. Furthermore, a directivity of the receiver could be specified, like microphone directivity or a head related transfer function. The result is an impulse response as it could exist for such a room.

The stochastic room acoustic simulation is no approach to model an exact room, but rather to find room impulse responses as they could exist in typical rooms. Using the same boundary conditions, a stochastic room acoustic simulation will return different impulse responses for every evaluation. This makes it rather useful for Monte-Carlo simulations. The method of stochastic room acoustic simulation is similar to the method of mirror sources. Therefore, it is better suited for frequencies above the 'Schroeder Frequency', although, due to time-frequency connections, it will also yield a realistic transfer function with discrete modes for lower frequencies (SCHROEDER, 1996).

The method of stochastic impulse responses is mostly based on a geometrical acoustics approach, as explained in (KUTTRUFF, 2000) and (VORLÄNDER, 2008).

The average temporal density of reflections arriving at time t in an arbitrary shaped room of the volume V can be expressed as:

$$\frac{\mathrm{d}N_\mathrm{r}}{\mathrm{d}t} = 4\pi \frac{c^3 t^2}{V} \qquad (2.39)$$

This means the density of sound reflections increases according to a quadratic law. The total number of reflections in an impulse response of the length t_ir is

$$N_\mathrm{ir} = \int_0^{t_\mathrm{ir}} 4\pi \frac{c^3 t^2}{V} \mathrm{d}t \qquad (2.40)$$

$$= \frac{4\pi}{3} \frac{c^3 t_\mathrm{ir}^3}{V} \qquad (2.41)$$

Those reflections should be distributed in the interval $]t_\mathrm{direct} \quad t_\mathrm{ir}]$, where t_direct is the time offset after which the direct sound reaches the receiver and prior to which no reflections can occur. A simple way of creating a random value with a quadratic distribution in the range $[0 \ t_\mathrm{ir}]$ is taking the cube root of an equal distribution P_equal, with $P_\mathrm{equal} \in [0 \ 1]$ and multiplying it with t_ir.

$$t_\mathrm{ps} = \sqrt[3]{P_\mathrm{equal}} \cdot t_\mathrm{ir} \qquad (2.42)$$

The temporal reflection distribution can be transferred to a distribution of image source distances.

$$d_\mathrm{ps} = c \cdot t_\mathrm{ps} \qquad (2.43)$$

The direction of the image sources relative to the head can be chosen randomly with an equal distribution on the full sphere. An equal distribution on a sphere can be generated by the following method:

$$z \in [-1 \ \ 1] \tag{2.44}$$

$$t \in [0 \ \ 2\pi] \tag{2.45}$$

$$r = \sqrt{1 - z^2} \tag{2.46}$$

$$x = r \cdot \cos(t) \tag{2.47}$$

$$y = r \cdot \sin(t) \tag{2.48}$$

With the combination of random distances and random directions, a set of image sources can be generated. The transfer function from image source i to the receiver with the directivity H_{receiver}, which depends on the direction of incidence in spherical coordinates ϕ and θ, can be calculated as:

$$H_i(\omega) = \frac{1}{ct_i} \cdot H_{\text{receiver}}(\omega, \phi, \theta) \cdot e^{\left[-j\omega - \frac{m(\omega)}{2}c + \bar{n}\ln(R(\omega))\right]t_i} \tag{2.49}$$

$$\bar{n} = \frac{cS}{4V} \tag{2.50}$$

with t_i being the delay between the source and the receiver, m the air attenuation, $\bar{n}$ the average number of reflections per second for one sound ray, and R the effective reflection factor. The values m and R will be frequency dependent for most situations.

The final transfer function is generated by a summation over all reflections and the direct sound, which is calculated by the same formulas with $n = 0$.

$$H = H_{\text{direct}} + \sum_{i=1}^{N_{\text{ir}}} H_i \tag{2.51}$$

Monte Carlo simulations

Monte Carlo simulations are a method of repeated random calculations to understand the connection of input and output parameters of a modeled problem. They are especially helpful when the problem includes a great number of coupled degrees of freedom. In acoustics, Monte Carlo simulations are often used to determine measurement uncertainties (MARTIN, WITEW, and ARANA, 2007).

2.4 Spatial Coherence

Coherence is a similarity indicator for signals in the frequency domain. It can also be seen as an indicator for the linearity and time invariance of a system. Identical signals have a coherence of unity. This is also true for all linear and time-invariant operations on one or both of the signals. Non-linear or time-variant parts in the transfer path decrease the coherence.

The coherence between two signals x and y with the power spectral densities S_{xx} and S_{yy} and the cross spectral density S_{xy} is defined as

$$\gamma_{xy}(\omega) = \frac{S_{xy}(\omega)}{\sqrt{S_{xx}(\omega) \cdot S_{yy}(\omega)}}. \tag{2.52}$$

A measurement over an infinite period of time period is impossible. Instead, a coherence estimation is performed by estimating the spectral densities using Fourier transformations of overlapping time blocks (CARTER, KNAPP, and NUTTALL, 1973)

$$\hat{\gamma}_{xy}(\omega) = \frac{\langle XY^* \rangle}{\sqrt{\langle XX^* \rangle \cdot \langle YY^* \rangle}} \tag{2.53}$$

where X and Y are the frequency domain representations of the signals x and y. The slanted brackets indicate averaging over time and/or frequency according to (MARPLE, 1987; WELCH, 1967). In this context, the block size n_{bs} of the segments is relevant, as the coherence estimate is biased by the block size (CARTER, KNAPP, and NUTTALL, 1973; SCHARRER and VORLÄNDER, 2010). Further details on the spectral density (SD) estimation in real-time systems are given in Section 2.5.1

The spatial coherence describes the coherence between two measures at two locations. The measures usually used in acoustics are the sound pressure p and the sound velocity $\boldsymbol{u}$.

The coherence between two pressure receivers is a complex, frequency dependent scalar:

$$\gamma_{p_1 p_2}(\omega) = \frac{\langle P_1 P_2^* \rangle}{\sqrt{\langle P_1 P_1^* \rangle \cdot \langle P_2 P_2^* \rangle}} \tag{2.54}$$

As the particle velocity $\boldsymbol{u}$ is a vector, the coherence between sound velocity and sound pressure is a vector parallel to the sound velocity:

$$\gamma_{pu}(\omega) = \frac{\langle P\boldsymbol{u}^* \rangle}{\sqrt{\langle PP^* \rangle \cdot \langle \boldsymbol{u} \cdot \boldsymbol{u}^* \rangle}} \tag{2.55}$$

2.4.1 Spatial Coherence in Basic Sound Fields

For the basic sound fields the spatial coherences can be deduced in an analytical fashion.

In free field conditions, the magnitude squared spatial coherence between two receivers, pressure or velocity, at any distance, is unity. The coherence function for any two acoustic sensors with a spacing d and an angle of sound incidence θ can be expressed as (KUTTRUFF, 2000)

$$\gamma_{xy}(\omega) = e^{-jkd\cos(\theta)} \tag{2.56}$$

Accordingly, the magnitude of γ_{xy} is always unity. The coherence estimate, however, is not always unity, as the result is biased by the time delay between the two receivers (CARTER, 1987). Just to give an example: One could chose the extreme where the time delay between the two receivers is bigger than the block size used for coherence estimation. In this case, the coherence would degrade to a value close to zero, even though the signals are in fact highly coherent. In less extreme cases, a time delay can bias the result of the coherence estimation as well. Carter (CARTER, 1987) gives an analytical estimation of the bias $E[\hat{C}] - C$ as a function of the true coherence C, the FFT time duration t_{bs}, and the time delay Δt

$$E[\hat{C}] - C \cong \frac{-2|\Delta t|}{t_{\text{bs}}} C + \left(\frac{|\Delta t|}{t_{\text{bs}}}\right)^2 C \tag{2.57}$$

This means that for a valid estimation the block size for the coherence estimation has to be chosen significantly higher than the maximum possible time delay between the two receivers.

PIERSOL (1978) shows that the magnitude squared spatial coherence between the signals of two pressure receivers in a diffuse sound field can be predicted as

$$\gamma_{xy}^2(k, d) = \left(\frac{\sin(kd)}{kd}\right)^2 \qquad (2.58)$$

where k is the wave number and d is the sensor spacing.

JACOBSEN and ROISIN (2000) extends this derivation to combinations of pressure and sound velocity sensors. The magnitude squared spatial coherence between a pressure and sound velocity signal in radial direction with a sensor distance d is

$$\gamma_{pu_r}^2(k, d) = 3\left(\frac{\sin(kd) - (kd)\cos(kd)}{(kd)^2}\right)^2 \qquad (2.59)$$

In the extreme case that pressure and velocity are measured at the same position, so that $d \to 0$, this is reduced to

$$\lim_{d \to 0} \gamma_{pu_r}^2(k, d) = 0 \qquad (2.60)$$

This means that in a perfectly diffuse sound field the sound pressure and velocity at the same position are perfectly incoherent.

The spatial coherence estimate in a reverberant sound field is similar to that in a diffuse sound field as long as the block size used for coherence estimation is small enough (JACOBSEN and ROISIN, 2000). A room with one stationary source is usually also considered both a linear and a time invariant system and those systems have a spatial coherence of unity. This is also valid as long as the block size for the coherence estimation is chosen significantly higher than the reverberation time of the room. This conflict also indicates that the result of the coherence estimation in a reverberant sound field is biased by the block size used for the coherence estimation.

In a reactive sound field, the magnitude of the spatial coherence is unity, no matter whether the spatial coherence between two pressure signals or a pressure and a velocity signal is evaluated (JACOBSEN, 1989). Like in free field conditions, the coherence can be biased by a time delay. This has to be considered if the time delay is not much smaller than the block size used for coherence estimation. The complex coherence between sound pressure and sound velocity at the same position has the same argument as the complex intensity at that same location (JACOBSEN, 1989).

The spatial coherence of two independent and uncorrelated noise signals is obviously zero, although, due to a limited averaging time and limited block size, a coherence estimate will return values above zero.

Figure 2.10 shows the spatial coherences for two pressure sensors with a sensor distance of 1 cm and the spatial coherence between pressure and velocity at the same location in the basic sound fields. Figure 2.11 shows the locations of the basic sound fields in a three-dimensional space, spun up by the spatial coherences $|\gamma_{pp}|$, $\Re\{\gamma_{pu}\}$ and $\Im\{\gamma_{pu}\}$.

Spatial coherence of combined signals

The spatial coherence γ_{xy} between two signals x and y, which are mixtures of N signals from different origins, can be predicted by the coherences of the single components x_i and y_i

$$x = \sum_{i=1}^{N} x_i \tag{2.61}$$

$$y = \sum_{i=1}^{N} y_i \tag{2.62}$$

The signal components are assumed to be uncorrelated to each other, so that in the frequency domain:

$$\langle X_a X_b^* \rangle \approx 0 \text{ for } a, b \in [1 \ N], a \neq b \tag{2.63}$$

$$\langle Y_a Y_b^* \rangle \approx 0 \text{ for } a, b \in [1 \ N], a \neq b \tag{2.64}$$

$$\langle X_a Y_b^* \rangle \approx 0 \text{ for } a, b \in [1 \ N], a \neq b \tag{2.65}$$

According to (2.53), the coherence for two signal components ($X(f) = X_1(f) + X_2(f)$ and $Y(f) = Y_1(f) + Y_2(f)$) can be expressed as:

$$\gamma_{xy} = \frac{\langle (X_1 + X_2)(Y_1 + Y_2)^* \rangle}{\sqrt{\langle (X_1 + X_2)(X_1 + X_2)^* \rangle \cdot \langle (Y_1 + Y_2)(Y_1 + Y_2)^* \rangle}} \tag{2.66}$$

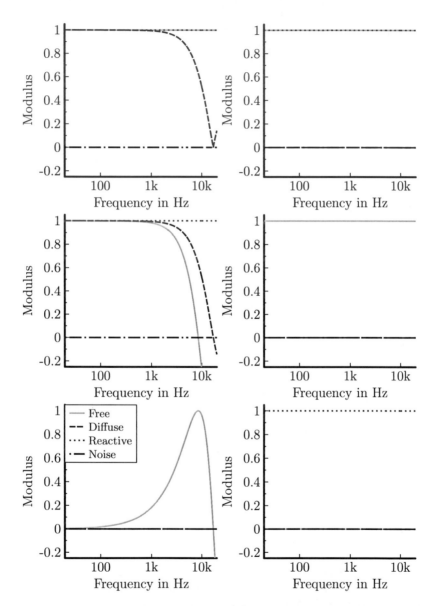

Figure 2.10: Magnitude (top), real (middle) and imaginary (bottom) parts of the spatial coherence for two pressure sensors with a sensor distance of 1 cm (left) and spatial coherence between pressure and velocity at the same location (right) in the basic sound fields

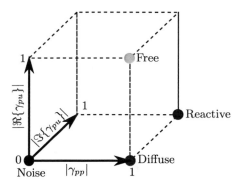

Figure 2.11: Locations of the basic sound fields in a three-dimensional space, spun up by the spatial coherences $|\gamma_{pp}|$, $\Re\{\gamma_{pu}\}$ and $\Im\{\gamma_{pu}\}$.

Using (2.63) to (2.65), this can be simplified to:

$$
\begin{aligned}
\gamma_{xy} \approx \quad & \frac{\langle X_1 Y_1^* \rangle}{\sqrt{\langle X_1 X_1^* \rangle \langle Y_1 Y_1^* \rangle}} \cdot \frac{\sqrt{\langle |X_1|^2 \rangle \langle |Y_1|^2 \rangle}}{\langle |X_1||Y_1| \rangle + \langle |X_2||Y_2| \rangle} \\
+ & \frac{\langle X_2 Y_2^* \rangle}{\sqrt{\langle X_2 X_2^* \rangle \langle Y_2 Y_2^* \rangle}} \cdot \frac{\sqrt{\langle |X_2|^2 \rangle \langle |Y_2|^2 \rangle}}{\langle |X_1||Y_1| \rangle + \langle |X_2||Y_2| \rangle}
\end{aligned}
\tag{2.67}
$$

Using (2.53), this can be expressed as a superposition of the coherences of the single signal components

$$
\begin{aligned}
= \quad & \gamma_{x_1 y_1} \cdot \frac{\sqrt{\langle |X_1|^2 \rangle \langle |Y_1|^2 \rangle}}{\langle |X_1||Y_1| \rangle + \langle |X_2||Y_2| \rangle} \\
+ & \gamma_{x_2 y_2} \cdot \frac{\sqrt{\langle |X_2|^2 \rangle \langle |Y_2|^2 \rangle}}{\langle |X_1||Y_1| \rangle + \langle |X_2||Y_2| \rangle}
\end{aligned}
\tag{2.68}
$$

And with the frequency dependent signal energies $E_x(f)$ and $E_y(f)$:

$$
E_x(f) = \langle |X|^2 \rangle
\tag{2.69}
$$

$$
E_x(f) = \sum_{i=1}^{N} E_{x_i}(f)
\tag{2.70}
$$

$$
E_y(f) = \sum_{i=1}^{N} E_{y_i}(f)
\tag{2.71}
$$

33

the frequency dependent coherence can be written as:

$$
\begin{aligned}
\gamma_{xy} = \quad & \gamma_{x_1 y_1} \cdot \frac{\sqrt{E_{x_1} E_{y_1}}}{\sqrt{E_{x_1} E_{y_1}} + \sqrt{E_{x_2} E_{y_2}}} \\
+ & \gamma_{x_2 y_2} \cdot \frac{\sqrt{E_{x_2} E_{y_2}}}{\sqrt{E_{x_1} E_{y_1}} + \sqrt{E_{x_2} E_{y_2}}}
\end{aligned}
\tag{2.72}
$$

With the assumption that the signal energies at both sensors are similar, so that

$$
E_{x_i} \approx E_{y_i} \tag{2.73}
$$

and

$$
\sqrt{E_{x_i} E_{y_i}} \approx E_{x_i} \tag{2.74}
$$

the equation can be written as

$$
\gamma_{xy} = \gamma_{x_1 y_1} \cdot \frac{E_{x_1}}{E_{x_1} + E_{x_2}} + \gamma_{x_2 y_2} \cdot \frac{E_{x_2}}{E_{x_1} + E_{x_2}} \tag{2.75}
$$

This means that the coherence of a combined signal can be predicted by the coherences of the single components with regard to their relative energies.

Equation (2.75) also corresponds to the equations derived for the relation between the SNR and the coherence by CARTER, KNAPP, and NUTTALL (1973), CARTER (1987) and JEUB (2012) and the relation between DRR and the coherence described by BLOOM and CAIN (1982), KUSTER (2011) and THIERGART, DEL GALDO, and HABETS (2012).

Equation (2.75) can be further generalized for the superposition of N uncorrelated signals:

$$
\gamma_{xy} \approx \sum_{i=1}^{N} \gamma_{x_i y_i} \cdot \frac{\sqrt{E_{x_i} E_{y_i}}}{\sqrt{E_x E_y}} \tag{2.76}
$$

$$
\approx \sum_{i=1}^{N} \gamma_{x_i y_i} \cdot \frac{E_{x_i}}{E_x} \tag{2.77}
$$

2.4.2 Coherence Estimate Function

For infinitely long signals, the coherence cannot be calculated but has to be estimated. For this purpose, the coherence is segmented into (partly overlapping) blocks of the length n_{bs} and the coherence is estimated by averaging over some of these blocks (for details see Section 2.5.1). In some cases, the block size used for the calculation of the spectral densities has a significant influence on the result of the coherence estimation (CARTER, 1987; CARTER, KNAPP, and NUTTALL, 1973). This effect is explained in literature, although there seems to be no general analytical approach to explain the amount of error due to the limited block size analysis. This block size dependency may also include some useful information on the system.

Therefore, the coherence estimate function (CEF) is defined as a function of the block size estimation, where $E(\gamma_{xy})|_{n_{bs}}$ indicates a coherence estimation with block size n_{bs}.

$$\text{CEF}(n_{bs}, f) = E(\gamma_{xy})|_{n_{bs}} \tag{2.78}$$

The CEF can be calculated as a mean value over the whole or a specific frequency range. As well, it can be calculated as individual representation for frequency bands like thirds, octaves, or Bark bands.

The block size n_{bs} can also be expressed as a time t_{bs}. The connection is the sampling rate f_s so that:

$$t_{bs} = \frac{n_{bs}}{f_c} \tag{2.79}$$

2.5 Sound Field Indicators

The term *sound field indicator* was first used in the context of sound intensity measurements by JACOBSEN (1989):

> A sound field indicator may be defined as a normalized quantity which describes some local or global property of the field. ...
> The theoretical examples demonstrate that it is possible to deduce quite useful information about the structure of the sound field from

the two sound field indicators suggested in Section 4.1, the normalized complex intensity and the coherence between pressure and particle velocity. Additional information can be deduced if the coherence function is measured (or computed) with variable frequency resolution.

A sound field indicator (SFI) is a property whose behavior is determined by the actual sound field the sensor(s) are exposed to, and not influenced by the source signal. The normalization ensures the independence from the source signal. Jacobsen proposes mainly two SFIs, the normalized complex intensity and the magnitude squared coherence between pressure and particle velocity. The SFIs are calculated from spectral densities that have to be estimated prior to the SFI calculation.

2.5.1 Spectral Density Estimation

As the input signals from a microphone array are infinite and possibly not stationary, the spectral densities need to be estimated. This progress is usually done by segmenting the input data into (overlapping) blocks of a constant number of samples (WELCH, 1967). The number of samples of each block is often called block size n_{bs}. For the final estimation of the auto and cross power SDs, an average over N segments is calculated.

$$S_{xx} = \frac{1}{N} \sum_{n=1}^{N} X_n(f)^* \cdot X_n(f) \tag{2.80}$$

$$S_{xy} = \frac{1}{N} \sum_{n=1}^{N} X_n(f)^* \cdot Y_n(f) \tag{2.81}$$

To calculate the immediate SD estimate at a block m, the last N blocks are averaged.

$$S_{xx,m} = \frac{1}{N} \sum_{n=m-N}^{m} X_n(f)^* \cdot X_n(f) \tag{2.82}$$

$$S_{xy,m} = \frac{1}{N} \sum_{n=m-N}^{m} X_n(f)^* \cdot Y_n(f) \tag{2.83}$$

An average over all segments is not feasible for infinite signals and the caching of many blocks is very memory intensive, and therefore not realistic for most real-time systems. Accordingly, an exponential average is often used in memory and time critical systems. The SD can then be calculated from the spectrum of the current block along with the result of the precedent block.

$$S_{xx,n} = \alpha_c \cdot (X_n(f)^* \cdot X_n(f)) + (1 - \alpha_c) \cdot S_{xx,n-1} \qquad (2.84)$$

The same method can be used to calculate the cross spectral density:

$$S_{xy,n} = \alpha_c \cdot (X_n(f)^* \cdot Y_n(f)) + (1 - \alpha_c) \cdot S_{xy,n-1} \qquad (2.85)$$

where α_c is an arbitrary number between zero and one that determines how strong the smoothing effect of the averaging should be. α_c can also be specified as a time constant t_c, similar to that of a first-order low pass. It depends on n_{bs}, an optional overlap of the blocks n_{ol} and the sampling rate f_{s}.

$$t_c = -\frac{1}{\ln(1 - \alpha_c)} \cdot \frac{n_{\mathrm{bs}} - n_{\mathrm{ol}}}{f_{\mathrm{s}}} \qquad (2.86)$$

$$\alpha_c = 1 - e^{\left(-\frac{1}{t_c} \cdot \frac{n_{\mathrm{bs}} - n_{\mathrm{ol}}}{f_{\mathrm{s}}}\right)} \qquad (2.87)$$

2.5.2 Calculation

All SFIs can be calculated from the spectral density estimation explained in Section 2.5.1. The calculation of the spatial coherence was already introduced in Section 2.4. The coherence is calculated from the cross SD normalized by the two signals' power spectral densitys (PSDs). In most cases, the coherence is presented as the magnitude squared coherence; for the purpose of the sound field classification, however, the complex coherence is necessary.

$$\gamma_{pp} = \frac{S_{p_1,p_2}}{\sqrt{S_{p_1,p_1} S_{p_2,p_2}}} \qquad (2.88)$$

$$\gamma_{pu} = \frac{S_{pu}}{\sqrt{S_{pp} S_{uu}}} \qquad (2.89)$$

The intensity is a measure of sound energy propagation. The instantaneous intensity can be calculated as $i = p \cdot u$ (FAHY, 1989). The instantaneous intensity is a vector directing in the same direction as the sound velocity u. In most intensity measurement applications, using two sound pressure probes or sound gradient probes, only the sound velocity in one normal direction can be evaluated; the instantaneous intensity is then a scalar, indicating only the component of the

37

intensity in the normal direction. The intensity is calculated from the instantaneous intensity by time averaging. Alternatively, the complex intensity i_c can be calculated in the frequency domain from the frequency domain presentations of the sound pressure P and sound velocity U as: (JACOBSEN, 1989)

$$i_c = \frac{1}{2}PU^* \tag{2.90}$$

with

$$P = \mathcal{F}\{p(t)\} \tag{2.91}$$
$$U = \mathcal{F}\{\boldsymbol{u}(t)\} \tag{2.92}$$

The normalized intensity is the complex intensity normalized by the magnitude squared sound pressure.

$$I_n = \frac{PU^*}{PP^*} \tag{2.93}$$

For non-stationary signals this can also be expressed using the spectral density estimations S_{pu} and S_{pp}

$$I_n = \frac{S_{pu}}{S_{pp}} \tag{2.94}$$

In free field conditions, the normalized intensity of a source straight ahead becomes one if using a normalized sound velocity or $\rho_0 c$ if using the physical correct sound velocity. As only the sound velocity in normal direction is evaluated in this case, the normalized intensity is also direction dependent so that $I_n = \cos(\theta)$. The components $I_{a,n} = \Re(I_n)$ and $I_{r,n} = \Im(I_n)$ indicate the active and reactive components of the normalized sound intensity.

The normalized impedance is basically the inverted normalized intensity. The calculation is accordingly:

$$Z_n = \frac{S_{pp}}{S_{pu}} \tag{2.95}$$

The normalized impedance does not yield any additional information on the sound field, as all information is already present in the normalized intensity.

The normalized transfer function is defined as the transfer function between the two pressure receivers of one hearing aid, normalized by the pressure at one sensor.

$$H_n = \frac{S_{p_1,p_2}}{S_{p_1,p_1}} \tag{2.96}$$

The SFIs have been extended in prior works like (STREICH, 2010; STREICH et al., 2010). The extended set of SFIs includes power spectral densities, cross power spectral densities, normalized front and back cardioid, as well as normalized fixed and adaptive beamformer. These extensions of the SFIs show some information for the purpose of situation classification (STREICH, 2010). The main disadvantage for the purpose of a sound field classification is that all SFI extensions depend on the source position or head orientation, the signal spectrum, or source strength. This means that they change, for example, when the microphone array is turned or the source signal changes in type or level, although this should not affect the type of sound field in the room, at least for situations with one dominant source.

Frequency bands

An evaluation of the sound field indicators in the spectral resolution that is returned by the (fast/discrete) Fourier transformation (FFT) for the SD calculation is typically not target-aimed, although possible. Usually, room acoustic parameters are given in octaves or thirds. Especially the last ones are supposed to represent the bands of the human hearing quite well (BLAUERT, 2005). DIN-EN-ISO-266 (1997) defines the calculation of octave and third middle frequencies (f_m), which also define the lower and upper frequency limits (f_l and f_u) of each band.

The SFI for each frequency band f_m can be calculated as a mean value from the SFI with a FFT frequency resolution Δf as

$$\text{SFI}_{x,f_m} = \frac{\Delta f}{f_u - f_l} \sum_{f=f_l}^{f_u} \text{SFI}_x(f) \tag{2.97}$$

The SFIs will be used for the sound field classification as described in Chapter 3. A thorough analysis of their behavior in the basic sound field is performed in Section 3.1.1.

3

Sound Field Classification

Parts of this chapter have been published in (SCHARRER and FELS, 2012), (SCHARRER and VORLANDER, 2013) and (SCHARRER and FELS, 2013)

A wide range of signal processing strategies are used in modern hearing aids and other mobile devices such as mobile phones to increase speech intelligibility, reduce noise and thus increase user satisfaction. Most of these strategies are beneficial in some but not all situations. If these strategies are not adjusted correctly, they might even disturb the user. Therefore, the performance of these signal processing strategies depends to a large extent on several boundary conditions like the signal itself or the sound field (ALEXANDRE et al., 2007).

Most multi-channel signal processing strategies are based on the assumption that a certain type of acoustic sound field is used. Beamforming methods postulate, for instance, a plane or spherical wave propagation in free field conditions, while noise cancellation methods in mobile devices depend on a diffuse noise field. In the case of mobile devices like hearing aids or mobile phones, different acoustic situations therefore require different signal processing strategies. Modern hearing aids often offer a set of predefined programs. These programs adapt the different stages of the signal processing to the individual situation (STREICH et al., 2010). The manual selection of the appropriate program by the user is a rather annoying and difficult task. An automatic program selection is highly appreciated by users and many modern hearing aids provide such a feature.

An automatic program selection requires a large number of features to classify the actual acoustic situation and user preferences. The target classes that are currently used in hearing aids are usually *speech, speech in noise, music* and *noise* (BÜCHLER et al., 2005). These classes describe the signal perceived at the device as a combination of the source signal(s), the room acoustic properties, and other influences like concurring sources or noise at the microphones due to electrical noise or external sources like wind. The concrete features that are evaluated for this purpose are essential for the results of the program selection (ALEXANDRE et al., 2007). Based on this observation, it is obvious that in addition to the established, signal-based sound classification, an additional set describing the acoustic sound field would be useful for the signal processing and automatic control strategies in mobile devices (HATZIANTONIOU and MOURJOPOULOS, 2004). The sound field classification delivers information on the sound field as perceived by the device. The classification, for example, indicates whether there is one nearby dominant source or many concurring ones. A nearby sound source like a mobile phone is indicated by strong reactive components. The proposed method is able to distinguish the occurrence of wind noise, which is classified as *noise*, from noise coming from a loudspeaker, which is composed of a *free* field with an additional *diffuse* component if reverberation is present.

Typical properties that influence the performance of common signal processing strategies in mobile devices are the SNR, the reverberation time, and the DRR (HAMACHER et al., 2005). The active-to-reactive-energy ratio (ARR) also influences some signal processing strategies. All these properties are a by-product of the sound field classification and can be received at no additional cost. The main advantage of the sound field classification, however, lies in the automatic program selection, whose performance can be increased significantly by using additional information on the sound field above signal properties alone (STREICH et al., 2010).

In the context of sound intensity measurements Jacobsen already discussed the possibility to differentiate between active and reactive as well as coherent and incoherent sound fields (JACOBSEN, 1989). In this context, he coined the term sound field indicator (SFI) as properties of the sound field that can be estimated alongside the intensity and that are normalized so that they describe local or global properties of the sound field. Accordingly, they are independent from the signal itself. In contrast to the classification method proposed in this thesis, the method by Jacobsen does not result in a sound field classification. The idea of the sound field indicators was expanded by STREICH et al. (2010) for the purpose of sound classification and reverberation time estimation. The authors extended the set of sound field indicators significantly, although, in contrast to Jacobsen's initial intention, most extensions to the sound field indicators are not independent from the source signal.

Recently GAUTHIER et al. (2011) introduced a method for a sound field classification for microphone arrays using a sound field extrapolation method. Three different classes are used to classify the sound field: *directive sound field, standing wave* and *diffuse sound field.* The method is only applicable to microphone arrays with a large number of sensors. In contrast, based on a single sensor input, NORDQVISTA and LEIJON (2004) discusses a 'listening environment classification' based on a hidden Markov model with the the classes *clean speech, speech in traffic noise* and *speech in babble.* This approach focuses more on classifying the background noise rather than the acoustic environment.

3.1 Problem Statement

The target of a sound field classification is to determine the energy distribution of the basic sound fields in a combined sound field. The aim is to classify the sound field rather than the signal. Accordingly, the sources' signals should not influence the classification result. The method should work in time variant situations, for example, when the receiver or a source moves. The method should also be able to run in real time, at least from the basic conception.

In contrast to the existing publications, the method described in this chapter is based solely on the input of two to four acoustic sensors. The method is independent from the source signal. Accordingly, it implements a sound field classification, rather than a signal classification. Furthermore, the results of the classification also reflect the energy distribution of the target classes. Some important characteristic values like the SNR or the DRR can be calculated directly from the results of the sound field classification. The classes meet the typical boundary conditions and assumptions required for the best performance of multi-channel signal processing strategies.

The basic assumption is that the sound field is a combination of four components. The total sound field energy E_{total} is then distributed among the four components

$$E_{\text{total}} = E_{\text{free}} + E_{\text{diffuse}} + E_{\text{reactive}} + E_{\text{noise}} \tag{3.1}$$

3.1.1 Input Signals and Sound Field Indicators

Due to the boundary conditions set for the sound field classification, the sound field indicators introduced in Section 2.5 were an promising starting point for a classification. The basic SFIs are signal independent and neither the source type nor the strength has an influence. The orientation of the sensor should not matter, too. Basically all SFI extensions described in Section 2.5.2 do not suit these criteria, so they were not evaluated in depth.

Table 3.1 shows the initial set of SFIs described in Section 2.5 and their expected behavior in the basic sound fields.

	Free	Diffuse	Reactive	Noise
$\|\gamma_{pu}\|$	1	0	1	0
$\angle(\gamma_{pu})$	0, π	random	$\frac{\pi}{2}$, $\frac{3\pi}{2}$	random
$\|\gamma_{pp}\|$	1	$\frac{\sin(kd)}{kd}$	1	0
$\angle(\gamma_{pp})$	$\omega d\cos(\theta)$	random	0, π	random
$\|I_n\|$	$\cos(\theta)$	0	>0	random
$\angle(I_n)$	0, π	random	$\frac{\pi}{2}$, $\frac{3\pi}{2}$	random
$\|Z_n\|$	$\frac{1}{\cos(\theta)}$	∞	<1	random
$\angle(Z_n)$	0, π	random	$\frac{\pi}{2}$, $\frac{3\pi}{2}$	random
$\|H_n\|$	1	$\frac{\sin(kd)}{kd}$	≈ 1	random
$\angle(H_n)$	$\omega d\cos(\theta)$	random	0, π	random

Table 3.1: Sound Field Indicators in Basic Sound Fields.

From a comparison of the SFIs in the different sound fields in Table 3.1 it is obvious that at least three of the indicators are necessary to discriminate the four basic sound fields. Random behavior of the SFIs in some of the basic sound fields is also not very useful in a classification context.

From Table 3.1 is also obvious that the complex coherence γ_{pu} and the magnitude of γ_{pp} are sufficient to separate all basic sound fields. They have the advantage that they are limited to the complex unity circle, and for all basic sound fields the magnitudes are known from analytic predictions.

Redundancies like using the normalized intensity along with the normalized impedance could yield some stability advantages if the values were calculated from different input values. As in this context they result from the same error prone input values, no information or stability gain is to be expected.

Further analysis will accordingly be based solely on the complex pu-coherence as well as on the magnitude of the monaural pp-coherence.

The estimation of these coherences puts some requirements on the input signals.

The pressure signals at two nearby points, as well as the sound pressure and the sound velocity at the same location, are necessary for the classification. Possible input forms are combinations of sound pressure and velocity sensors, sound field microphones or microphone arrays, as long as the necessary input values can be measured or calculated from the available sensors.

Two examples, of a hearing aid setup and a sound field microphone, will describe how the input signal can be gathered from small microphone arrays.

Hearing aids

In a setup with two pressure sensors with the signals p_1 and p_2, the other input signals can be calculated as:

$$P = \frac{P_1 + P_2}{2} \tag{3.2}$$

$$U_r = \frac{P_2 - P_1}{jkd} \tag{3.3}$$

This is a valid approximation as long as the sensor spacing d is much smaller than the wavelength $\lambda = \frac{2\pi}{k}$.

This will, however, lead to a direction-dependent term for a sound source with an angle θ from the normal direction, so that

$$U_r = ||\boldsymbol{u}|| \cdot \cos(\theta) \tag{3.4}$$

where U_r the is the measured sound velocity and $\boldsymbol{u}$ is the actual sound velocity vector.

This does not matter in case of single coherences, as the direction-dependent term is canceled out in (2.55). It does however matter for the superposition of signals. This will be explained in detail in Section 3.2.2.

Sound field microphone

For a small microphone array with four pressure sensors arranged in a regular tetrahedron shape with an edge length a and the pressure input signals P_x

$$\boldsymbol{p}_{\text{in}} = \begin{bmatrix} P_a & P_b & P_c & P_d \end{bmatrix}^{\text{T}} \tag{3.5}$$

the required input for the sound field classification can be calculated as

$$P = \frac{1}{4} \begin{bmatrix} 1 & 1 & 1 & 1 \end{bmatrix} \cdot \boldsymbol{p}_{\text{in}} \tag{3.6}$$

$$\boldsymbol{u} = \frac{\sqrt{2}Z_0}{jka} \begin{bmatrix} 1 & 1 & -1 & -1 \\ 1 & -1 & -1 & 1 \\ 1 & -1 & 1 & -1 \end{bmatrix} \cdot \boldsymbol{p}_{\text{in}} \tag{3.7}$$

where Z_0 is the characteristic impedance of air. If d is small enough, so that (2.58) returns $\gamma_{xy}^2(k,d) \approx 1$, the microphone input signals can be used directly for the calculation of γ_{pp}, so that, for example, $P_1 = P_a$ and $P_2 = P_b$. Otherwise, the arbitrary distance d' should be chosen small, so that the condition is fulfilled. γ_{pp} can then be calculated from

$$P_1 = P + \frac{jkd'}{\sqrt{2}Z_0} \begin{bmatrix} 1 & 0 & 0 \end{bmatrix} \cdot \boldsymbol{u} \tag{3.8}$$

$$P_2 = P - \frac{jkd'}{\sqrt{2}Z_0} \begin{bmatrix} -1 & 0 & 0 \end{bmatrix} \cdot \boldsymbol{u} \tag{3.9}$$

3.1.2 Target Classes

The basic sound fields, as already explained in Section 2.2.1, were chosen as target classes for the sound field classification: *free*, *diffuse*, *reactive* and *noise*. They also represent the most common sound fields described in room acoustic literature, such as (FAHY, 1989; JACOBSEN, 1989; KUTTRUFF, 2000). Adding the class *noise*, these target classes also concur with the classes proposed for a sound field classification by GAUTHIER et al. (2011).

3.2 Classification

Based on the spatial coherences and their known behavior for the four target classes described in Section 2.4.1, a new sound field classification approach is introduced. Apart from a strict classification, which indicates only the dominant sound field, knowledge about the energy distribution between the different sound field components is obtained. For the purpose of the classification, two approaches are introduced. First, a classical fuzzy distance classification was used. This method proved to be very robust to sensor errors und other influences. The fuzzy distance classification returns only an approximate energy distribution; even in best cases the classification is never perfect. The second approach is based on inverting the equation system of the coherence superposition. It therefore theoretically returns perfect results. The method is, however, constrained by a strong influence of sensors errors and other influences, and behaves unstable in a way that it can even return negative energies for the single classes.

3.2.1 Distance Classification Approach

A fuzzy distance classification can be used to reconstruct the sound field composition from the feature vector.

The feature vector SFD consists of three features as follows:

$$\text{SFD} = \begin{pmatrix} |\gamma_{pp}| \\ |\Re(\gamma_{pu})| \\ |\Im(\gamma_{pu})| \end{pmatrix} \tag{3.10}$$

The locations of the target classes in this feature space can be derived from room and signal processing literature as:

$$\text{SFD}_{\text{free}} = \begin{pmatrix} 1 \\ 1 \\ 0 \end{pmatrix} \qquad \text{SFD}_{\text{diffuse}} = \begin{pmatrix} \left| \frac{\sin(kd)}{kd} \right| \\ 0 \\ 0 \end{pmatrix}$$

$$\text{SFD}_{\text{reactive}} = \begin{pmatrix} 1 \\ 0 \\ 1 \end{pmatrix} \qquad \text{SFD}_{\text{noise}} = \begin{pmatrix} 0 \\ 0 \\ 0 \end{pmatrix}$$

The classification is performed by calculating the distance d_c between the feature vector and the target classes for every time index n

$$d_c(n) = \|\text{SFD}(n) - \text{SFD}_c\| \tag{3.11}$$

To determine the membership values μ_c, each class has to be assigned a weight $w_c(n)$. This is usually done by inverting the Euclidean distance. In this case, a different approach was chosen. Using the energy relation between SNR and the resulting coherence from (CARTER, KNAPP, and NUTTALL, 1973), the energy distribution between the classes is estimated as

$$w_c(n) = \frac{1}{d_c(n)} - 1 \tag{3.12}$$

Finally, the membership values $\mu_c \in [0, 1]$ can be calculated from the weights by normalizing them to a sum of one (NOCK and NIELSEN, 2006):

$$\mu_c = \frac{w_c(n)}{\sum_i w_i(n)} \tag{3.13}$$

The total energy of each sound field can be reconstructed by a multiplication of the membership values with an estimation of the total energy.

$$\begin{pmatrix} E_{\text{free}} \\ E_{\text{diffuse}} \\ E_{\text{reactive}} \\ E_{\text{noise}} \end{pmatrix} = \mu \cdot E_{\text{total}} \tag{3.14}$$

3.2.2 Unified Approach

Assuming that an actual sound field is composed of the four basic sound fields, and the signals of the single components are uncorrelated to each other, at least within the block size used for coherence estimation, one can use (2.77) for a superposition of the spatial coherences of all four basic sound fields. That means the general spatial coherence of the superposition of all four basic sound fields can be expressed as

$$\gamma_{xy} = \gamma_{xy}|_{\text{free}} \cdot \frac{E_{\text{free}}}{E_{\text{total}}} + \gamma_{xy}|_{\text{diffuse}} \cdot \frac{E_{\text{diffuse}}}{E_{\text{total}}}$$
$$+ \gamma_{xy}|_{\text{reactive}} \cdot \frac{E_{\text{reactive}}}{E_{\text{total}}} + \gamma_{xy}|_{\text{noise}} \cdot \frac{E_{\text{noise}}}{E_{\text{total}}} \tag{3.15}$$

where $E = \sqrt{E_p \cdot E_u}$. As the normalized sound velocity is evaluated the energies E_p and E_u are similar, the simplification in (2.74) is valid.

Along with the knowledge about the theoretic spatial coherences in the basic sound fields, and using normalized energies $E' = \frac{E}{E_{\text{total}}}$, and assuming that all normalized energies have to sum up to one, this can be expressed for the spatial coherences used for the sound field classification:

$$\begin{pmatrix} \gamma_{\text{pp}} \\ \Re\{\gamma_{\text{pu}}\} \\ \Im\{\gamma_{\text{pu}}\} \\ 1 \end{pmatrix}^{\text{T}} = A \cdot \begin{pmatrix} E'_{\text{free}} \\ E'_{\text{diffuse}} \\ E'_{\text{reactive}} \\ E'_{\text{noise}} \end{pmatrix} \tag{3.16}$$

with

$$A = \begin{pmatrix} 1 & \frac{\sin(kd)}{kd} & 1 & 0 \\ 1 & 0 & 0 & 0 \\ 0 & 0 & 1 & 0 \\ 1 & 1 & 1 & 1 \end{pmatrix} \tag{3.17}$$

This equation can be inverted to obtain the normalized energies of the basic sound fields from the spatial coherences:

$$\begin{pmatrix} E'_{\text{free}} \\ E'_{\text{diffuse}} \\ E'_{\text{reactive}} \\ E'_{\text{noise}} \end{pmatrix} = A^{-1} \begin{pmatrix} \gamma_{\text{pp}} \\ \Re\{\gamma_{\text{pu}}\} \\ \Im\{\gamma_{\text{pu}}\} \\ 1 \end{pmatrix}^{\text{T}} \tag{3.18}$$

with

$$A^{-1} = \begin{pmatrix} 0 & 1 & 0 & 0 \\ \frac{kd}{\sin(kd)} & -\frac{kd}{\sin(kd)} & -\frac{kd}{\sin(kd)} & 0 \\ 0 & 0 & 1 & 0 \\ -\frac{kd}{\sin(kd)} & -\frac{(\sin(kd)-kd)}{\sin(kd)} & -\frac{(\sin(kd)-kd)}{\sin(kd)} & 1 \end{pmatrix} \tag{3.19}$$

Furthermore, the total energy of each sound field can be reconstructed by a multiplication of the relative energy with an estimation of the total energy.

$$
\begin{pmatrix} E_{\text{free}} \\ E_{\text{diffuse}} \\ E_{\text{reactive}} \\ E_{\text{noise}} \end{pmatrix} = F_{\text{total}} \cdot A^{-1} \cdot \begin{pmatrix} \gamma_{\text{pp}} \\ \Re\{\gamma_{\text{pu}}\} \\ \Im\{\gamma_{\text{pu}}\} \\ 1 \end{pmatrix}^{\text{T}}
\tag{3.20}
$$

For the case of a two microphone input, the energies perceived at the sensor become direction-dependent (see (3.4)), so that (2.74) is no longer valid and A needs to be rewritten to take this into consideration:

$$
\begin{pmatrix} \gamma_{\text{pp}} \\ \Re\{\gamma_{\text{pu}}\} \\ \Im\{\gamma_{\text{pu}}\} \\ 1 \end{pmatrix}^{\text{T}} = A \cdot \begin{pmatrix} E'_{\text{free}} \\ E'_{\text{diffuse}} \\ E'_{\text{reactive}} \\ E'_{\text{noise}} \end{pmatrix}
\tag{3.21}
$$

$$
A = \begin{pmatrix} 1 & \frac{\sin(kd)}{kd} & 1 & 0 \\ \cos(\theta) & 0 & 0 & 0 \\ 0 & 0 & \cos(\theta) & 0 \\ \cos(\theta) & 1 & \cos(\theta) & 1 \end{pmatrix}
\tag{3.22}
$$

Accordingly, A^{-1} for the sound field classification changes to:

$$
A^{-1} = \begin{pmatrix} 0 & \frac{1}{\cos(\theta)} & 0 & 0 \\ \frac{kd}{\sin(kd)} & -\frac{kd}{\cos(\theta)\sin(kd)} & -\frac{kd}{\cos(\theta)\sin(kd)} & 0 \\ 0 & 0 & \frac{1}{\cos(\theta)} & 0 \\ -\frac{kd}{\sin(kd)} & \frac{kd-\cos(\theta)\sin(kd)}{\cos(\theta)\sin(kd)} & \frac{kd-\cos(\theta)\sin(kd)}{\cos(\theta)\sin(kd)} & 1 \end{pmatrix}
\tag{3.23}
$$

For the sound field classification, this means that the direction of the source(s) has to be known a priori. Depending on the application, a beamforming algorithm, a binaural auditory model, or other methods to gather this knowledge could be utilized.

3.3 Verification

Two different approaches were chosen to verify of the sound field classification. The first approach consists of an audio sequence of simulated basic sound fields. The second approach compares the results of the sound field classification from a room acoustic simulation to those of a theoretic prediction of the sound field energies.

3.3.1 Basic Sound Fields

For a basic verification of the sound field classification, the BSFS described in Section 2.3.1 was classified. The signals received at the different sensors were generated using analytic models as described in Section 2.3.1. The sequence included the basic sound fields in the sequence *free*, *diffuse*, *reactive* and *noise*. It was expected that each part of the sequence be classified accordingly. The simulated microphone array consisted of four microphones in a tetrahedron alignment. The spacing between the microphones was set to 12 mm. The classification was performed with a frequency resolution of one third octave bands. The block size used for the coherence estimation was set to 256 samples with an overlap of 192 samples at a sampling rate of 44.1 kHz. t_c was set to 0.5 s.

The broadband classification result for both classification methods is shown in Figure 3.1. The BSFS is also indicated to illustrate the ideal classification results. Both methods are able to classify all four basic sound fields in the correct sequence. The *free* and *reactive* sound fields are classified without any obvious error. The classification of the sound fields *diffuse* and *noise* shows a slight error. This fault occurs due to the limited block size and number of blocks used for the coherence estimation. Even for uncorrelated noise, a slight correlation is to be expected when using finite block sizes. Figure 3.2 shows the frequency-dependent classification results. These results indicate that the method works quite well in the whole frequency range that is evaluated. For high frequencies, the sound velocity estimation produces some errors that reflect as slightly reactive in the classification of the *free* sound field. For low frequencies, the microphone spacing is too small for a valid velocity estimation, a problem that is already known in sound intensity probes that utilize a similar sensor setup. The errors in the sound velocity estimation lead to classification errors for low frequencies. For high frequencies, the spatial coherence in a diffuse sound field decreases. The coherence in this case depends on the sensor spacing. As the coherence decreases, the diffuse sound field cannot be distinguished from the case of *noise*. Accordingly, the sensor spacing influences performance contrarily at high and low frequencies. An optimal sensor spacing depends on the frequency range of interest. The change from one sound field to another in the simulated sound sequence is immediate, so the coherence estimates at the intersections are defective. This leads to classification errors at the transfer from one sound field to another. This also means that abrupt changes in the sound field will lead to classification errors. This adaption time is determined by the time constant t_c used for the spectral density estimation.

3.3.2 Room Acoustic Theory

A second verification was performed using a room acoustic simulation of a rectangular room, a setup of a point source and four microphones aligned in a tetrahedron alignment. The simulated room had a reverberation time of 1 s. The simulation method used is an adapted mirror source model, details can be found in Section 2.3.2 and (SCHARRER and VORLÄNDER, 2010). The sensor alignment allows for a calculation of the sound velocity vector as well as the sound pressures required for the sound field classification. The block size used for the coherence estimation was set to 256 samples with an overlap of 192 samples at a sampling rate of 44.1 kHz. t_c was set to 1 s. Using this setup, multiple distances between the source and the receivers were simulated and the results of the classification were compared with those from theoretic predictions described in (KUTTRUFF, 2000). The layout of the experiment is indicated in Figure 3.3. The results for the octave band with the center frequency of 1 kHz and the classification approach are shown in Figure 3.4. The results for the unified approach are shown in Figure 3.5. The distance r between point source and receiver is normalized by the critical distance r_c. The results of the sound field classification were multiplied by the total sound field energy to estimate the absolute sound field energies of each class. The sound field energy was normalized by the energy at the critical distance.

For the dominant sound field, the sound field classification using the classification approach is very close to the theoretic predictions. Close to the source, the diffuse field energy is underestimated slightly. For distances bigger than the critical distance, the energies of the direct and reactive sound field are overestimated. The relative energy estimation shows bigger deviations than the absolute estimation, which results from the logarithmic scale of the absolute energy estimation, which conceals some of the deviations. The relative energy estimation shows that close

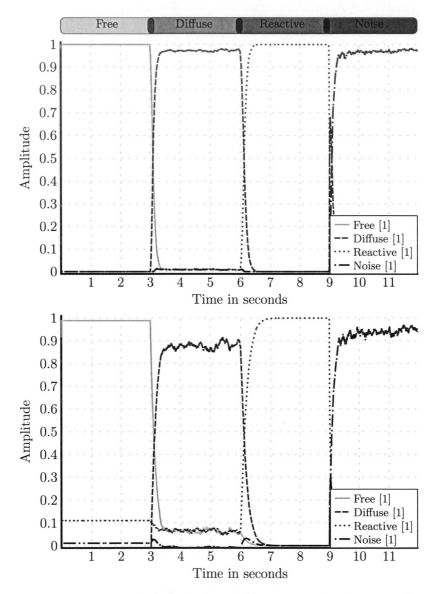

Figure 3.1: Broadband classification of the BSFS, using the classification approach (top) and the unified approach (bottom).

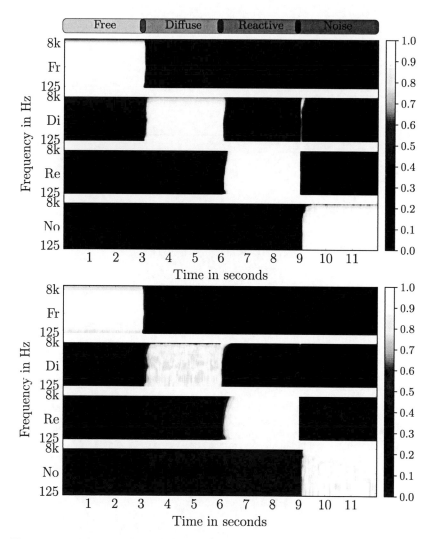

Figure 3.2: Frequency-dependent classification of the BSFS, using the classification approach (top) and the unified approach (bottom). From top to bottom: Classification as *free*, *diffuse*, *reactive* and *noise*.

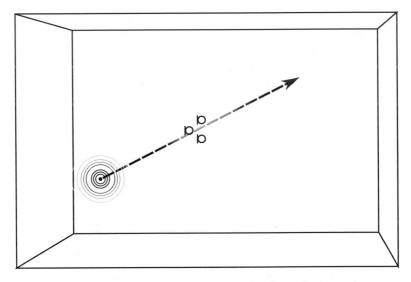

Figure 3.3: Layout for an SFC verification experiment. A microphone array and a point source are placed in a room. The microphone array is moved away from the point source in a straight line. The sound field classification in dependence of the distance to the point source is calculated and compared to theoretic predictions.

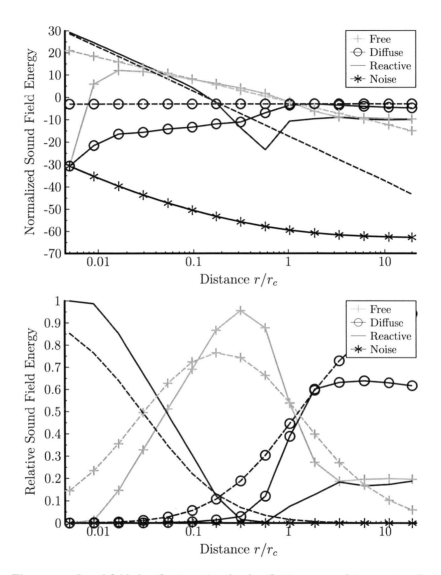

Figure 3.4: Sound field classification using the classification approach in a rectangular room as a function of the distance between source and receiver. Classification results are solid lines, predictions from diffuse field and point source theory are dotted. Top: absolute energy estimation. Bottom: relative energy estimation

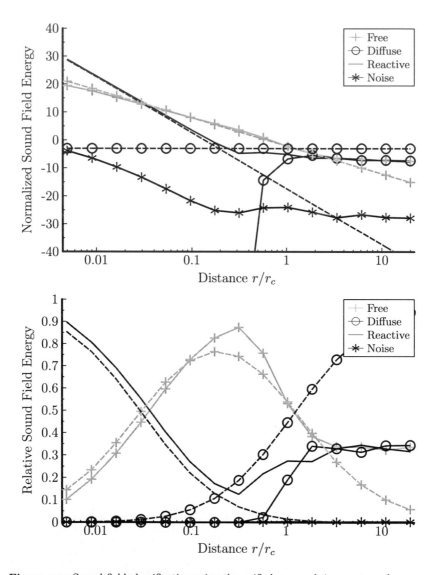

Figure 3.5: Sound field classification using the unified approach in a rectangular room as a function of the distance between source and receiver. Classification results are solid lines, predictions from diffuse field and point source theory are dotted. Top: absolute energy estimation. Bottom: relative energy estimation

to the source the *reactive* energy is overestimated. A little further away from the source, the *free* energy is overestimated. In the far field, the *diffuse* energy is underestimated, and the *free* as well as the *reactive* energies are overestimated.

The unified approach also shows very precise results for the estimation of the absolute energy of the dominant sound field. For distances below the critical distance, the diffuse sound field energy is underestimated significantly. For distances above the critical distance, the sound fields *free*, *diffuse* and *diffuse* are estimated with similar energies, which underestimates the *diffuse* sound field and overestimates the other two. The estimation of relative sound field energies is much more precise than the classification approach. For distances above r_c the estimation error from the absolute estimation is reflected.

For underlying sound fields with an energy lower than -10 dB compared to the dominant sound field, the classification using both methods is not very precise. For distances above the critical distance, the diffuse sound field energy is underestimated and the energies of free and reactive sound fields are overestimated. For all distances some noise is estimated, although the simulation was noise-free. The estimation of noise as well as the underestimation of the diffuse sound field energy result from errors in the coherence estimation due to limited averaging time and block size.

Figure 3.6 shows an evaluation of the unified approach in a scene with additional sensor noise with the same level as the diffuse sound field. The evaluation was performed with two different time constants. In the first experiment, the time constant was chosen as $t_c = 1$ s. In the second experiment the time constant is adjusted to $t_c = 10$ s. The classification performance increases, especially for the inferior sound fields. Although this increase of the time constant returns significantly better classification, results it will most probably not be suitable for applications in mobile devices, as the corresponding adaption to changes in the acoustic environment is rather slow and may result in significant classification errors in time variant situations.

3.4 Error Sources

The following section shall determine the robustness of the developed sound field classification sensor errors, such as microphone mismatch.

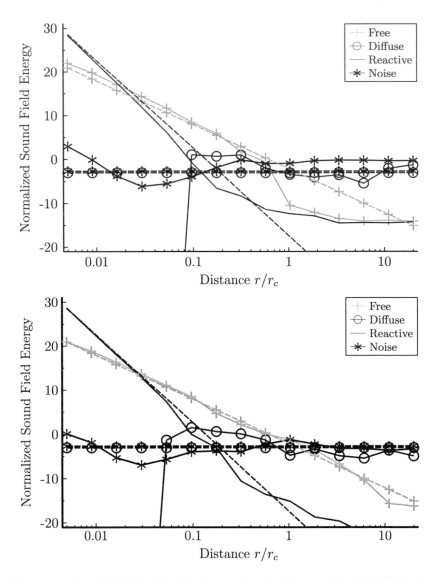

Figure 3.6: Sound field classification in a rectangular room as a function of the distance between source and receiver, normalized by the critical distance r_c. Classification results are solid lines, predictions from diffuse field and point source theory are dashed. (Top: $t_c = 1\,\text{s}$, Bottom: $t_c = 10\,\text{s}$)

Typical sensor errors are amplitude and/or phase errors that lead to a mismatch between the two sensors. Sensor noise is another typical problem occurring when dealing with real sensors. The absolute magnitudes of the frequency dependent sensitivities of the used sensors are not important, as they do not influence the results of the sound field classification. Depending on the sensor setup, some values need to be estimated, such as the sound velocity, if a microphone array is used as input for the classification. In this case, the estimation of the sound velocity deals with the same problems as already encountered in sound intensity measurements. In this context, a difference in microphone sensitivities as well as differences in the phase response can lead to significant errors in the sound velocity estimation and the corresponding sound intensity estimation. The problem has been discussed by FAHY (1989), JACOBSEN (1991), JACOBSEN and BREE (2005), MOSCHIONI, SAGGIN, and TARABINI (2008) and many others. In the context of mobile devices, the problem is even more significant, as the sensors used are not expensive measurement sensors with low tolerances and low drift due to aging, but rather very cheap mass production elements from a big range of standard factory models with possibly significant drift due to aging.

No representative values on the typical sensor mismatch in applications like hearing aids could be established jet. The sensor mismatch of the four hearing aids used for the hearing aid dummy head are shown in Figure 3.7. The top graph shows the magnitude mismatch and indicates that a maximum value in amplitude of 1 dB is realistic. The middle graph shows the phase error of the microphones under test. The bottom graph shows the estimated frequency delay. The main reason for phase errors are different resonance frequencies of the microphones.

As a conclusion, from this rather small test group of hearing aid microphones, a maximal amplitude mismatch of 1 dB and a typical group delay of 0.1 ms seem realistic for reliability tests of the sound field classification.

For the evaluation of the influence of these sensors errors, the following experiment was carried out. The sound field classification needs at least two input signals, either that of two pressure sensors or that of a pressure sensor and a sound velocity or pressure gradient sensor. Accordingly, two minimum setups were evaluated. The first setup consists of two pressure sensors (pressure-pressure (PP)-setup). In the second case, a combination of a pressure sensor and a one-dimensional sound velocity sensor is evaluated (pressure-velocity (PU)-setup).

In a first step, a given situation with a known result from a sound field classification with two perfect sensors is evaluated. This result is the reference. From this

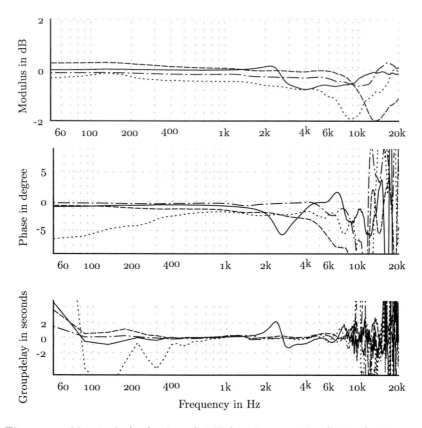

Figure 3.7: Magnitude (top), phase (middle) and group delay (bottom) differences between the two microphones of each of the four hearing aids under test.

point, an artificial mismatch is introduced to the sensors and the classification is reevaluated. The result of the repeated classification is then compared to the reference. The absolute deviation of the classification results, averaged over the time of the sound sample, is used as indicator for the influence of the sensor mismatch to the SFC.

Figure 3.8 shows the mean absolute error averaged over time for the classification of the BSFS, depending on sensor mismatch. Displayed are both sensor types and two types of sensor errors, a magnitude mismatch and a phase delay between the two sensors. In the case of a PP-setup, small amplitude mismatches as well as small phase delays lead to significant errors in the classification. The PU-setup seems more robust to sensor errors. An amplitude mismatch between the two sensors has only little influence. A phase delay shows some error, but smaller than for the PP-setup.

As the BSFS is quite artificial and may not represent the influences in a combined signal, a second test with a signal recorded in a reverberant room was carried out. The result of this second evaluation is shown in Figure 3.9. The results are very similar to those from the BSFS.

3.4.1 Automatic Sensor Mismatch Compensation

The errors in the sound field classification are quite serious, even for small sensor mismatches. Accordingly, when used with cheap uncalibrated sensors with a high sensitivity range from mass production models, no trustworthy results can be expected. For using the SFC in such a context, some form of calibration or compensation is necessary.

A number of automatic sensor mismatch compensation methods for sound intensity probes are discussed in literature. A broad overview on the topic is given in (JACOBSEN, 1991). All methods need a calibration process of some form. Often the sensors need to be put into free field conditions, a Kundt's tube or each sensor has to be calibrated on its own in a pressure chamber. This is practicable in measurement instruments like sound intensity probes, but not very applicable for mobile devices produced in mass production. Calibrating the sensors while manufacturing the product is complicated and increases costs. It also does not ensure against sensor drift due to aging, (partial) covering of the sensors with wind shields, hair or other changing effects.

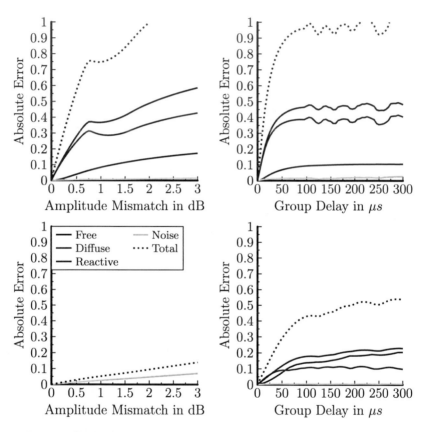

Figure 3.8: Absolute error averaged over time in the sound field classification of the BSFS, depending on sensor mismatch. Left: Amplitude mismatch; Right: Delay; Top: PP-setup; Bottom: PU-setup

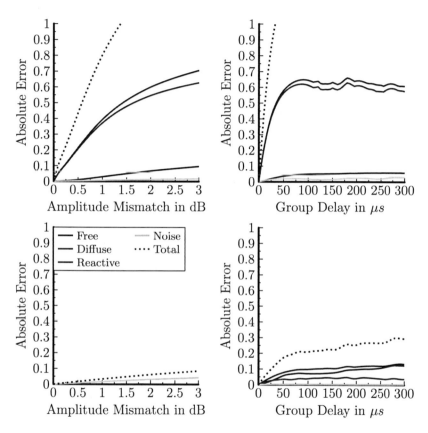

Figure 3.9: Absolute error averaged over time in the sound field classification of a speech in reverberation signal, depending on sensor mismatch. Left: Amplitude mismatch; Right: Delay; Top: PP-setup; Bottom: PU-setup

In combination with a sound field classification, a very simple method for the automatic compensation of a small sensor mismatch is possible. Whenever the sound field classification indicates a dominant diffuse sound field, one can assume that the two pressure signals should have the same energy per band as well as the same average phase per band. Using this knowledge, the sensor sensitivity mismatch can be determined and an average phase delay can be estimated. The estimated mismatch can then be used to compensate the error and achieve better input signals for the further evaluation. A slow adaption ensures against errors and makes the automatic compensation robust.

Figures 3.10 and 3.11 show the same evaluations of the mean absolute error averaged over time in the classification of the BSFS and speech signal as already shown in Figures 3.8 and 3.9 but with activated automatic sensor mismatch compensation. It seems that nearly all influences can be perfectly compensated so that finally the classification performance is not influenced by a sensor mismatch at all.

As the sensor mismatch itself decreases the performance of the sound field classification, it must not be so drastic that the sound field classification fails completely. In the simulation of the sensor influence this is taken care of by slowly increasing the error while the compensation adapts. That also means, that a significant sensor mismatch has to be trained at some point from which slow changes like sensor aging are adapted automatically. The first training could be achieved by placing the sensor array in a (partly) diffuse sound field and forcing an adaption.

3.5 Examples

A set of experiments was carried out using a dummy head equipped with two hearing aids and the proposed sound field classification. The hearing aid shell had a typical hearing aid housing and two microphones with a distance of 14 mm. The signals of the microphones were recorded directly, so the hearing aids did not process any signals. The recorded signals were then fed into the sound field classification, which performed a classification with a frequency resolution of one third octave bands. The block size used for the coherence estimation was set to 256 samples with an overlap of 192 samples at a sampling rate of 44.1 kHz. t_c was set to 0.5 s. The direction of the sound source was chosen as straight ahead and provided as a priori knowledge into the classification as necessary for this kind of

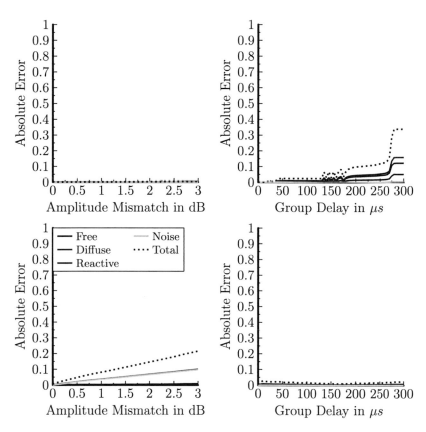

Figure 3.10: Absolute error averaged over time in the sound field classification of the BSFS, depending on sensor mismatch, using automatic sensor mismatch compensation. Left: Amplitude mismatch; Right: Delay; Top: PP-setup; Bottom: PU-setup

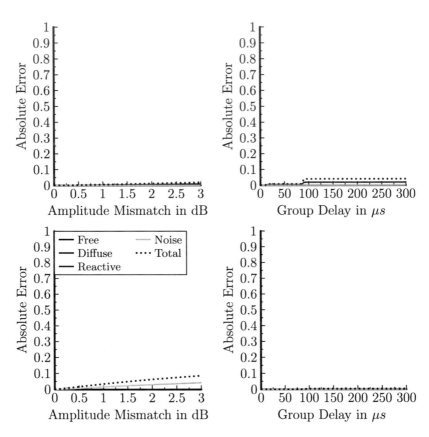

Figure 3.11: Absolute error averaged over time in the sound field classification of a speech in reverberation signal, depending on sensor mismatch, using automatic sensor mismatch compensation. Left: Amplitude mismatch; Right: Delay; Top: PP-setup; Bottom: PU-setup

microphone setup.

The spectrogram of the speech signal used in some of the examples is shown in Figure 3.12. The text, which is spoken by a woman, reads as follows:

> In language, infinitely many words can be written with a small set of letters. In arithmetic, infinitely many numbers can be composed from just a few digits, with the help of the symbol zero, the principle of position and the concept of a base. Pure systems with base five and six are said to be very rare. But base twenty occurs ...

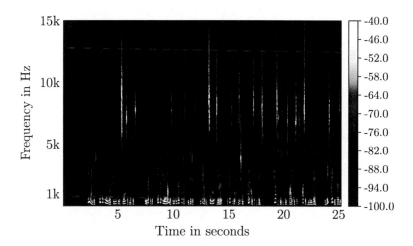

Figure 3.12: Spectrogram of the speech signal used for the SFC examples

Figure 3.13 shows the sound field classification of a quiet anechoic semi room. The sound field is classified as mainly *noise*, which seems accurate, as the only signal recorded is the electronic noise from the microphones and equipment like preamplifiers and audio devices. In a next step, a loudspeaker was placed in the anechoic semi room at a distance of about 2 m. The loudspeaker played a speech signal. The sound field in this case is classified as mostly *free* as shown in Figure 3.14, which is to be considered correct in an anechoic semi room. There are also some slightly reactive components, which could result from the floor reflection as well as a slight microphone mismatch, as the near field effect of the loudspeaker should not reach as far as 2 m.

Using the same setup, some recordings were performed in a reverberant room with a reverberation time of about 0.9 s. For the first setup, the loudspeaker was placed at a distance of about 1 m from the dummy head. The results of the sound field classification, shown in Figure 3.15, show that this distance is below the critical distance of the room. The sound field is mostly classified as *free* with some *diffuse* components from the reverberation. The diffuse components occur mostly during the pauses of the speech and vanish almost completely when words are spoken. Figure 3.16 shows the sound field classification of another setup in the same room but with an increased distance between the dummy head and the loudspeaker. In this case, the distance is bigger than the critical distance; the sound field is classified as *diffuse*.

Figure 3.17 shows the result of a sound field classification of wind noise. A rough recording of wind noise was optioned by recording a signal while slightly blowing over the hearing aid dummy. The wind noise is classified as *noise*.

3.6 Discussion

A method for the classification of a sound field in setups with multiple acoustic sensors was proposed and evaluated. The basic idea of describing sound fields by *sound field indicators* was developed by JACOBSEN (1989). The concept of a classification of the sound field based on those SFIs has not been described in scientific literature. The method can be applied, for example, in mobile devices with multiple acoustic sensors as well as in microphone arrays or sound intensity measurement applications. The aim of the classification is to categorize four basic sound fields *free, diffuse, reactive,* and *noise*. These classes also represent the most common boundary conditions for multi-channel signal processing and are well established assumptions for sound fields in rooms.

Based on the SFIs, two different approaches for the classification and estimation of sound field energies were evaluated. A classic distance-based classification in a feature space with known target class locations was tested, along with a unified approach that is based on a prediction of the spatial coherences of the superposition of the single sound fields. Both methods show similar but not identical results in most cases.

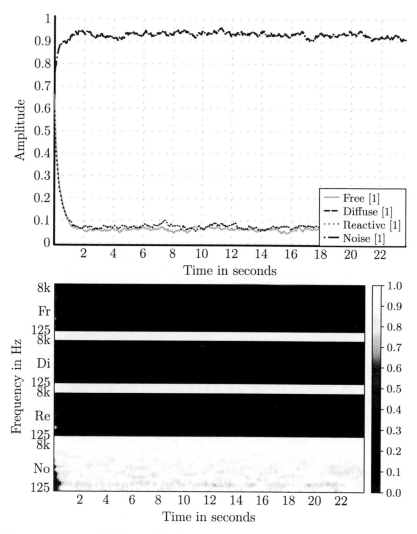

Figure 3.13: Sound field classification (using the unified approach) in a quiet semi anechoic room. From top to bottom: Classification as *free, diffuse, reactive* and *noise.*

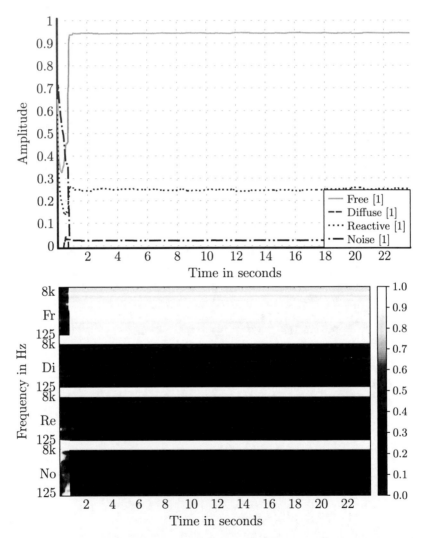

Figure 3.14: Sound field classification (using the unified approach) of a nearby speaker playing a speech signal in an anechoic semi room. From top to bottom: Classification as *free*, *diffuse*, *reactive* and *noise*.

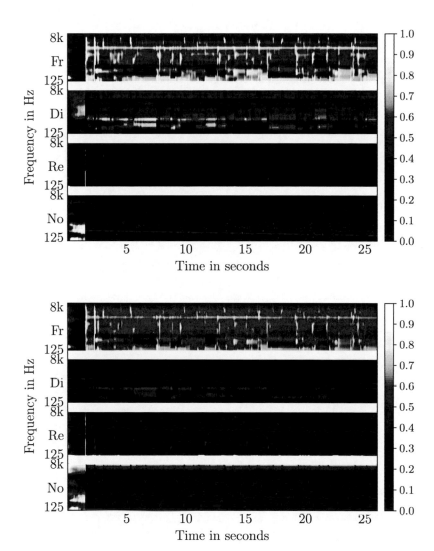

Figure 3.15: Sound field classification of a nearby speaker playing a speech signal in a reverberant room. Top graph: Classification approach, Bottom graph: Unified approach. From top to bottom: Classification as *free*, *diffuse*, *reactive* and *noise*.

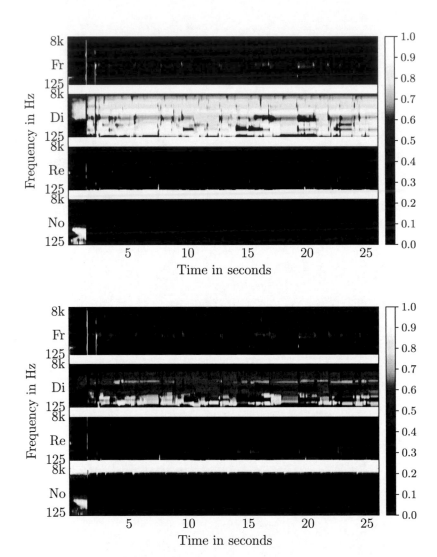

Figure 3.16: Sound field classification of a distant speaker playing a speech signal in a reverberant room. Top graph: Classification approach, Bottom graph: Unified approach. From top to bottom: Classification as *free, diffuse, reactive* and *noise*.

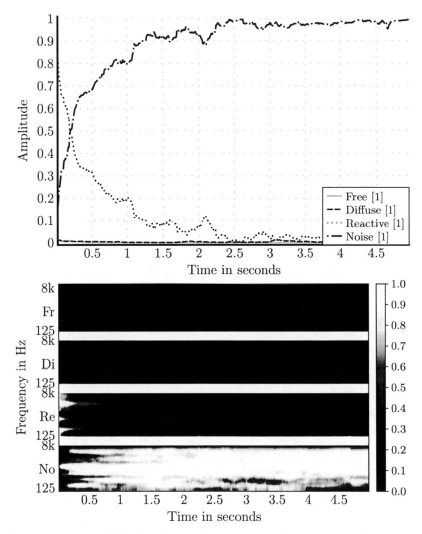

Figure 3.17: Sound field classification (using the classification approach) of a slight blow over the hearing aid. From top to bottom: Classification as *free*, *diffuse*, *reactive* and *noise*.

The classification method works independently of the source signal, as spatial coherences are the only indicators used for the classification. Noise from a speaker in free field conditions is classified as *free*, whereas microphone noise and wind-noise are classified as *noise*. The method is thus an extension of the sound classification already used in some hearing aids and is aimed at an application in the classification and control system of mobile devices that adapt the signal processing to the actual situation.

The method with both classifiers was verified with a sequence of simulated pure forms of the target classes. Both classification methods delivered a correct filing of the different sound fields to their classes. The methods were also verified by comparing the distance-dependent composition of the sound field in a reverberant sound field excited by a point source. The results show that both classifiers deliver a valid estimation of the dominant sound field while subsequent sound fields may show significant estimation errors. One factor for the precision of the methods is the time constant t_c used in the coherence estimation. A higher t_c delivers better classification results but slows the adaption in time variant situations. Both verification methods showed the general validity of the proposed sound field classification.

An approach for hearing aids with two microphones was also proposed. As two microphones are not sufficient for a classification, the result in this case is biased by the source position, as the sound velocity can only be estimated in the viewing direction of the hearing aid.

The robustness of the method against sensor mismatch was evaluated. This experiment showed a high influence of sensor mismatch on the sound field classification. Therefore, an automatic sensor mismatch compensation was proposed. The automatic compensation is able to completely compensate a sensor mismatch, as long as the mismatch is not so big that the results from the classification are completely wrong. Slow changes in sensor sensitivity due to aging can be fully compensated after an initial training.

A set of real life examples was evaluated, using hearing aid shells on a dummy head. All results showed plausible results.

A real-time capable demonstrator verifies the general applicability of the method in any real-time application. Other limitations, such as computation performance and energy consumption, are, however, not taken into account.

4

Blind Reverberation Time Estimation

Parts of this chapter have been published in (SCHARRER and VOR-LÄNDER, 2010) and (SCHARRER and VORLÄNDER, 2011)

The development of the sound field classification described in Chapter 3 leads to some discoveries about spatial coherence in acoustic sound fields. Especially reverberant sound fields show the behavior that the result of a coherence estimation depends on the analysis parameters used. This behavior is described in literature, for example by JACOBSEN and ROISIN (2000). This leads to the definition of the coherence estimate function (CEF), which describes exactly this dependency. After some experiments it seemed obvious that there is at least some connection between the CEF and some room acoustic properties like the reverberation time. As the CEF can be achieved without any knowledge about the source signal, it is an interesting indicator for the blind estimation of the reverberation time as necessary in mobile devices and their signal processing methods.

State-of-the-art hearing aids, and other audio processing instruments, implement signal processing strategies tailored to the specific listening environments. These instruments are expected to have the ability to evaluate the characteristics of the environment and accordingly use the most appropriate signal processing strategy (ALLEGRO, BÜCHLER, and LAUNER, 2001). Hence, a robust and reliable method to estimate the reverberation time from passively received microphone signals represents an important technology to improve the device's performance and the user experience.

Measurements and estimations of reverberation times have been the subject of many studies over the years. Measurements of reverberation time usually work with switched-off noise (SABINE, 1922) or impulse response measurements (SCHROEDER, 1965). The measurement procedure is standardized in ISO-3382-2 (2008). The reverberation time is an important and commonly quoted objective acoustic parameter for rooms. Reverberation influences speech intelligibility as well as music enjoyment. Reverberation also has a big influence on signal processing strategies, such as beam forming, time delay estimation or noise suppression. Therefore, knowledge about the reverberation time can improve the quality of the results of such signal processing. The main problem in this context is the determination of the reverberation time with given limits, like uncontrolled excitation and an unknown acoustic environment. In many situations no controlled excitation is possible. This starts with occupied rooms where people in the room would perceive the measurement sound, typically a sweep or noise, as annoying. Furthermore, for many applications no active excitation is possible at all – for example, in transportable devices that do not include a speaker. In those cases, only a blind reverberation time estimation, with no knowledge of the excitation signal itself, is possible. Especially, the blind estimation of reverberation times is still a field with a great deal of uncertainty and room for improvement. Most methods only work for special conditions, as they often make certain assumptions on the unknown excitation signal or the room.

Most methods for reverberation time estimation try to emulate the method of switched-off noise. In this case, a noise source excites a steady sound field in a room. After the noise source is switched off, the sound level in the room will decay linearly. An evaluation of this decay reveals the reverberation time (KUTTRUFF, 2000). The only difference for (semi blind) reverberation time estimation is that there is no control over the sound source. Some methods scan the audio signal for gaps and the level decay is evaluated (VESA and HARMA, 2005). Other methods are maximum likelihood estimation (RATNAM et al., 2003; ZHANG et al., 2006), neural networks (COX, LI, and DARLINGTON, 2001) or blind source separation (WANG, SANEI, and CHAMBERS, 2005). LOPEZ, GRENIER, and BOURMEYSTER (2012) use a decay rate distribution of the evaluated signals as indicator for the estimation.

Blind reverberation time estimation methods based on a source separation approach use the room impulse response as a by-product that can directly be evaluated, for example, by using the methods described in ISO-3382-2 (2008). However, this method has a critical drawback. It only works when the room impulse response is at minimum phase, a condition that is not met in most cases (RATNAM et al., 2003). Therefore, the method will not work in most environments.

Maximum likelihood methods usually try to estimate the reverberation time using the decay of the envelope of the autocorrelation function. Most of these methods have problems dealing with noise (LÖLLMANN and VARY, 2008), or coupled rooms, where the level of decay shows multiple decay rates (KENDRICK et al., 2007).

LÖLLMANN and VARY (2009) describe a single-channel noise reduction with a reverberation time estimation using a maximum likelihood approach. GAUBITCH et al. (2012) performed a comparison of different single-channel reverberation time estimation methods, one based on a maximum likelihood estimation of the reverberation tail slope, one based on modulation energy ratios, and one based on speech slope distributions. Under good conditions with an SNR greater than 30 dB, all methods are able to estimate the correct reverberation time within 0.2 s for short reverberation times below 0.8 s

There are also approaches using neural networks, which are trained with input signals for different rooms, although this method is not really blind as the network has to be trained with a known sound sequence. Reverberation time can only be estimated at occurrences of this sequence (COX, LI, and DARLINGTON, 2001).

A major drawback is that all those methods can be fooled by using an excitation signal with reverberation. This leads to a signal showing two decay rates, that of the room as well as that of the reverberated signal itself. Accordingly, most methods will return bad estimations. Additionally, the methods that can deal with multiple decay rates will return the combined reverberation time of signal and room, where e.g. the reverberation time of the room alone is critical for dereverberation methods and the reverberation in the source signal is unimportant and maybe even wanted.

4.1 Problem Statement

The following chapter shall verify that it is possible to estimate the reverberation time based on the spatial coherence and the CEF. The necessary conditions for this are similar to those of the sound field classification. The estimation can be based on the spatial coherence between sound pressure and sound velocity (PU-coherence) or the spatial coherence between two pressure receivers (PP-coherence). The estimation should be able to perform in real time, at least theoretically.

4.1.1 Influence of the Reverberation Time and other Parameters on the Coherence Estimate Function

The CEF between two acoustic receivers in a room is influenced by the reverberation time as well as the SNR and DRR. This section will try to determine the possible exactness of the estimation of the reverberation time T and DRR by using Monte Carlo simulations. In the first part, four different situations are evaluated. The room dimensions are constant with $5 \times 4 \times 3$ m. Two reverberation times are evaluated: 0.5 s and 1 s. The DRR was chosen as -10 dB and 10 dB. For each of those four situations, 100 different room impulse responses were simulated, using the stochastic room impulse response simulation described in Section 2.3.2. This method produces impulse responses that are not directly related to the room geometry, but to the room volume V, the surface S and the reverberation time T. The method produces stochastic impulse responses, so that every execution with the same settings returns a new and different impulse response. All impulse responses were convolved with white noise and the CEF was calculated for every impulse response.

Spatial pressure CEF (pp-CEF)

The CEF for spatial distributed pressure sensors was used in (SCHARRER and VORLÄNDER, 2010) for the estimation of the reverberation time. The results of a Monte Carlo simulation were used to determine the theoretical limitations of this reverberation time estimation method.

Figure 4.1 shows the CEF for the four situations and and 100 different stochastic simulation results. The receivers were located 20 cm apart from each other, as they would be for example in a binaural hearing aid setup. The CEF is not identical for all rooms. Instead, there are slight differences, which are lower for high block sizes and higher for the small block sizes. As the differences are most significant for low block sizes, it seems that especially the energy distribution and directions of the first reflections are the cause for the non-identical CEF.

CEF for sound pressure and sound velocity (pu-CEF)

The CEF between sound pressure and sound velocity behaves similarly to the CEF for two spatially distributed sensors. In free field conditions it is unity. In

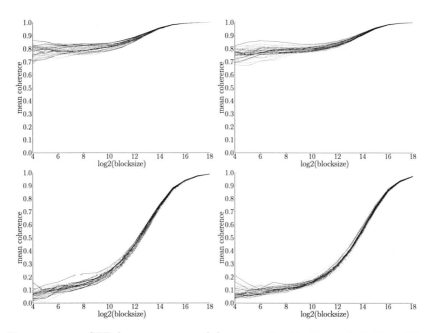

Figure 4.1: pp-CEF for 100 rooms and four acoustic situations. Left: $T = 0.5\,\mathrm{s}$, Right: $T = 1\,\mathrm{s}$, Top: DRR = 10 dB, Bottom: DRR = -10 dB

diffuse sound fields the coherence between the sound pressure and one component of the sound velocity, with a distance r between both measurement points, can be expressed as (JACOBSEN and ROISIN, 2000):

$$\gamma_{pu_x}^2(\omega, r) = 3 \left(\frac{\sin(\omega r/c) - (\omega r/c)\cos(\omega r/c)}{(\omega r/c)^2} \right)^2 \tag{4.1}$$

If sound pressure and velocity are evaluated at the same position, so that $r \to 0$, this equation converges towards zero.

Figure 4.2 shows the calculated CEFs values, averaged over the whole frequency range of the CEF for the simulated four situations and 100 rooms.

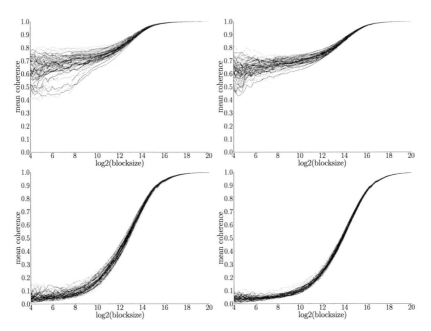

Figure 4.2: pu-CEF for 100 rooms and four acoustic situations. Left: $T = 0.5\,\mathrm{s}$, Right: $T = 1\,\mathrm{s}$, Top: DRR $= 5\,\mathrm{dB}$, Bottom: DRR $= \text{-}10\,\mathrm{dB}$

Influence of the reverberation time

All four situations in Figures 4.1 and 4.2 are clearly separated; a further experiment was conducted to determine the limits of a possible reverberation time estimation method. Simulations were performed with reverberation times between 0.2 s and 1.5 s with steps of 0.1 s. All simulations had the same DRR of -10 dB. For every reverberation time, 50 simulations were performed. Figure 4.3 shows the mean values and standard deviations of the resulting CEF. The influence of the reverberation time is visible in the medium block sizes as an increase of the reverberation time shifts the ascent of the CEF to higher block sizes. The standard deviations do not overlap in a medium block size range, leading to the assumption that a resolution of 0.1 s seems feasible for the estimation of the reverberation time.

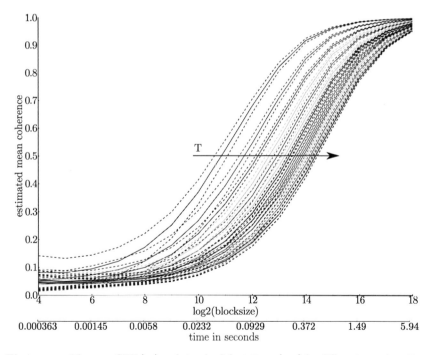

Figure 4.3: Mean pu-CEF (—) and standard deviations (- - -) for different reverberation times between 0.2 s and 1.5 s with a resolution of 0.1 s

Influence of the DRR

Another set of simulations shall determine the influence of the DRR on the CEF. Figure 4.4 shows the CEF in a room with a reverberation time of 1 s and a varying DRR. For a DRR of 0 dB and below, the CEF is almost zero for low block sizes. With the increasing DRR, the CEF also increases. Due to the big standard deviations, there is an overlap between the different scenarios. According to this figure, a resolution of less than 5 dB in the estimation of the DRR from the CEF is probably unrealistic.

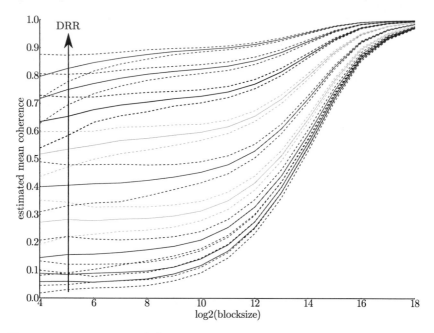

Figure 4.4: Mean pu-CEF (—) and standard deviations (- - -) for different DRR levels between −10 dB and 10 dB with a resolution of 2.5 dB

Influence of the SNR

Similar to the influence of the DRR, the influence of the SNR was determined by using a single room setup with a reverberation time of 1 s and a DRR of -10 dB. Using this setup, uncorrelated noise was added to all input channels. The results for different SNRs are shown in Figure 4.5. For an SNR of -10 dB, the CEF is close to zero in the whole block size range. With increasing SNR, the CEF for big block sizes increases, reaching 1 at an SNR of about 25 dB. This means, that the SNR range between -10 dB and 25 dB should be estimable from the CEF. The standard deviation of the CEF for high block sized is very small, so that a high resolution of under 1 dB for the SNR estimation seems feasible in the specified range.

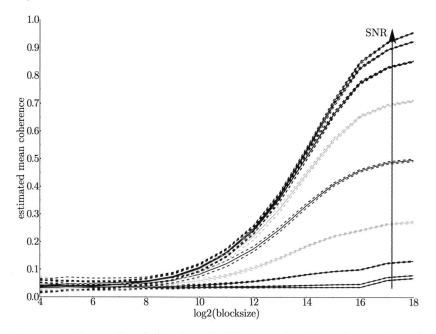

Figure 4.5: Mean pu-CEF (—) and standard deviations (- - -) for different SNR levels between -10 dB and 10 dB with a resolution of 2.5 dB

CEF frequency stability

Another important question is whether or not the CEF behaves similarly for all frequencies. For this purpose, an impulse response with a reverberation time that is constant over the whole frequency range was convolved with noise. The CEF was calculated for third octave bands. The result is shown in Figure 4.6. Obviously, the result is not the same for all frequency bands, but the trends are similar and the deviations are not too big.

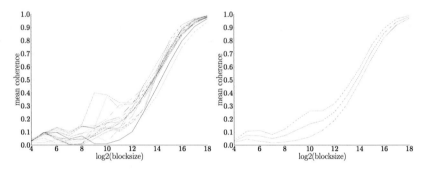

Figure 4.6: Left: pu-CEF for third octave bands; Right: Mean pu-CEF (—) and standard deviations (- - -)

Conclusion CEF influence factors

In conclusion, the CEF seems to be mostly dominated by the factors T, DRR and SNR as quantitatively indicated in Figure 4.7.

4.1.2 Input Signals

As discussed in the previous section, either the CEF between two spatial distributed pressure sensors (PP) or the CEF between sound pressure and sound velocity (PU) can be evaluated. The method depends on the differences between the coherences in direct (free) and diffuse sound fields. For the case of two spatial distributed pressure sensors, the sensors need to be placed with some significant spacing, as the coherence of a diffuse noise field (see (2.58)) depends on the frequency and the sensor spacing. For an evaluation, for example at low

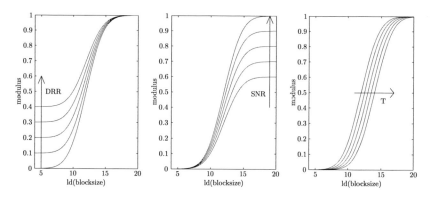

Figure 4.7: Quantitave influences of DRR, SNR and T on the CEF (SCHMOCH, 2011)

frequencies, the distance between the two ears of the human head is not sufficient. The coherence between sound pressure and sound velocity in a diffuse noise field at the same location is zero. Therefore, it is advantageous to measure them at the same location or at least very close to each other.

Depending on the sensor setup, either the PP-CEF or the PU-CEF is suited better for the blind reverberation time estimation. For example, for a small sensor array such as a sound field microphone, the PU-CEF is suited better, as no big sensor spacings are available, but the sound velocity and pressure at the center of the array can be estimated easily. By contrast, in a big microphone array like an acoustic camera, a valid estimation of the sound velocity is more ambiguous, even though spatially distributed sound pressure measurements are easily available.

4.2 Reverberation Time Estimation

The previous set of experiments showed that a reverberation time estimation based on the CEF seems feasible. For this purpose, three different approaches have been researched in the question how to get the reverberation time from a CEF. The first approach was to use a neural network. The main advantage here is that no basic knowledge about the underlying dependency between CEF and the reverberation time is necessary. Basically, if it is possible to train the network with a set of examples, and afterwards get reliable results from other examples not in the training data, there must be some dependency between the CEF and the reverberation time. The neural network approach was used as a proof of concept prior to more detailed research. The next step was to empirically model the CEF as a function of the reverberation time (and other parameters). This step still does not yield a deeper understanding of the underlying reasons. The method showed, however, what exact influence factors are resembled in the CEF. The final step was to model the CEF based on physical and system theoretical predictions. If this approach were successful, a formula connecting the CEF to T, SNR and DRR would be expected to yield superior estimation results. A simple inversion of the formula, a curve fit or maximum likelihood approach would return the estimated reverberation time.

4.2.1 Coherence Estimation and the CEF in Time-Variant Situations

The first crucial step in the reverberation time estimation is the CEF calculation. This can be done for different types of input signals, spacial pressure distributions, or sound pressure and sound velocity. The CEF has to be calculated according to (2.78) as indicated in Figure 4.8.

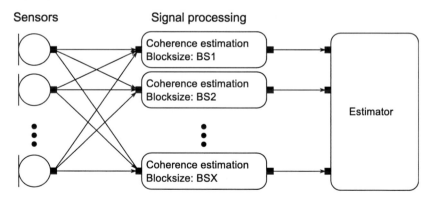

Figure 4.8: Scheme of the CEF calculation. The coherence is estimated for all combinations of input signals and for different block sizes n_{bs}. The results can then be fed into the reverberation time estimator.

For this purpose, the spectral densities of the input signals have to be estimated at different block sizes. Similar to the sound field classification, this is done by exponentially averaging the Fourier transformations of succeeding blocks of the input signals. In this case, one set of spectral densities has to be kept in memory for every block size evaluated. Accordingly, the SDs calculate as

$$S_{xy,n,n_{bs}} = \alpha_{n_{bs}} \cdot (X_{n,n_{bs}}(f)^* \cdot Y_{n,n_{bs}}(f)) + (1 - \alpha_{n_{bs}}) \cdot S_{xy,n-1,n_{bs}} \qquad (4.2)$$

The choice of $\alpha_{n_{bs}}$ is important, as it influences the behavior of the coherence in time variant situations. It is desired that SDs estimation for all block sizes behaves similar to changes of the transfer path. $\alpha_{n_{bs}}$ is calculated from a time constant t_c that can be chosen as suitable for the purpose of the sound field classification.

$$\alpha_{n_{bs}} = 1 - e^{\left(-\frac{1}{t_c} \cdot \frac{n_{bs} - n_{ol}}{f_s}\right)} \qquad (4.3)$$

Accordingly, $\alpha_{n_{bs}}$ is different for every n_{bs} but the SD estimation behaves similar.

The process of generating the input data (the CEF) for the estimation process is shown in Figure 4.8. Boundary conditions were set for the subsequent testing of the methods as well as the following examples. The block sizes used were in powers of two, with a step size of one half, in the range between 2^4 samples and 2^{20} samples. The signals used had a sampling rate of 44.1 kHz, so that the block sizes concur with time constants in the range of 0.3 ms to 23.8 s. The time constant t_c for the coherence estimation was set to 20 s. The output of the coherence calculations yielded 32 frequency-dependent values. The SD estimation with different block sizes also led to different frequency resolutions in the results. For the evaluation of the CEF, all results were converted to 512 frequency bins. This was achieved by linear interpolation for block sizes below 2^9 samples and reduction by averaging for block sizes above 2^9 samples.

The CEF resulting from those calculations was fed into the estimation process itself. Three different estimators were evaluated.

4.2.2 Estimation Using a Neural Network

The first estimator used was a neural network. This approach delivered fast results and was used to evaluate whether or not there was any possibility for a reliable estimation of the reverberation time from the CEF.

Neural networks

Artificial neural networks are designed as an abstraction of the signal processing in the human brain. A neural network consists of connected neurons. Neural networks are often used for pattern recognition and function interpolation. Usually they are designed as adaptive systems that change their structure based on information that is presented during a learning phase. They are very suitable for modeling complex or non-linear systems, with the advantage that the complex model dependencies may be unknown. Additionally, neural networks are insensitive to single errors in the input data, making the process itself robust (KRIESEL, 2007).

Estimator layout

A simple feed-forward network was chosen as estimator, including two hidden layers with 20 and 10 neurons per layer.

For frequency-dependent estimations there are two possibilities: either one network for every frequency of interest, or one network with one input for every frequency and block size. The second method has the advantage that typical dependencies between the frequencies could be considered by the neural network. For this first approach, only broadband values have been used for training and evaluation. All simulated rooms had reverberation times that were frequency independent.

Training data

The network had to be trained once. This could be done with a set of measured or simulated input data and the corresponding training targets. The training process had to be done for every microphone distribution. A simple way to generate sufficient amounts of training data are Monte Carlo simulations (see Section 2.3.2).

To limit the number of necessary simulations, the boundary conditions of the Monte Carlo simulations have been chosen as follows: The room dimensions were chosen randomly in the range between 1 m and 200 m; the reverberation time was chosen randomly using a normal distribution with a mean value according to

DIN-18041 (2004). The DIN-18041 (2004) describes proposed reverberation times for rooms with a given volume and purpose. It is meant as a guide in the acoustic layout of rooms for performances, education and other purposes. As a result, there was a slight correlation between reverberation time and room volume, as is to be expected for real rooms (KUSTER, 2008). Nevertheless, there can be rooms with the same volume and different reverberation times and vice versa. The distance between source and receiver was selected randomly, limited by the room geometry. The number of sources was limited to one. The source signal was chosen randomly from a pool of signals that included noise, which was created individually each time, and a set of sound files that included speed as well as music. Using this random description, the impulse responses could be calculated with the stochastic room acoustic model described in 2.3.2. Afterwards, the final receiver signal could be calculated by convolving the source signal with the impulse response, and then adding incoherent noise if a non-perfect SNR ratio is desired.

Results

Using the Monte Carlo simulation method, a set of 4000 different binaural signals was created, using the previously described boundary conditions. These signals were used to train and evaluate the neural network. Of the 4000 different rooms, 2000 were used for training, 1000 for verification of training results, and 1000 for a final test. For every situation, the binaural coherences for all block sizes were pre-calculated and saved for fast access so that it was not necessary to calculate the coherences for every training step.

The training was applied using a Levenberg-Marquardt backpropagation approach (LEVENBERG, 1944; MARQUARDT, 1963). This means that the training process was supervised and repeated until the gradient of the network performance undercut a certain threshold, meaning that only small improvements were to be gained by further training. The network performance is shown in Figure 4.9. For every training epoch, the relative mean squared error for the set of training, evaluation and test samples is shown. After 21 training epochs, the minimal error in the validation and test samples is reached. The final relative squared error for the training samples is about 10^{-5}, the one for the validation and test samples about 10^{-3}.

Figure 4.10 shows the results from the training, verification and final test. In each figure, every training sample is marked as a dot, indicating the neural networks output versus the real reverberation time, which is the training target. The samples used for the final test have not been included in any training process, so they are new to the neural network, whereas the other samples have all been presented in each training epoch. For all three training stages there is a very high correlation between neural network output and training target. The solid line indicates a linear least-squared error fit of the training results, the dotted line indicates the ideal result; training result equal to training target.

The neural network adapts very well to the presented training data and is able to model the relation between the CEF and the reverberation time. The relative error in all three sets is below 1 %, indicating a unique relation between the CEF and reverberation time.

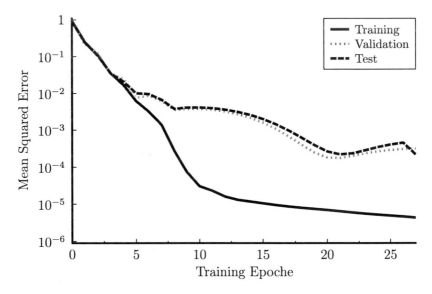

Figure 4.9: Relative mean squared error for the training epochs, separated into training, validation and test sets

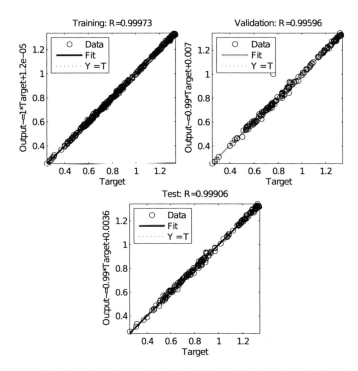

Figure 4.10: Results after the training (top left), validation (top right) and verification (bottom) of the neural network. The neural networks output is shown versus the training target. Each circle indicates a training sample. Additionally, a linear least-squared error fit and the ideal fit are indicated.

4.2.3 Empirical Approach

As long as no analytical solution for a model of the CEF was available, an empirical model seemed the next best solution. The basic idea to estimate the reverberation time from a CEF goes as follows:

1. Calculate the CEF from the input signals

2. Fit the modeled CEF_m to the calculated one

3. This results in one or more parameters

4. At least one of these parameters should have an obvious dependency on the reverberation time

5. From this parameter the reverberation time can be calculated or estimated.

SCHMOCH (2011) modeled the CEF based on an empirical approach. Due to the similarity between the CEF and the error-function erf (4.5), this function was chosen as basis for the model. To fit the erf to the CEF, it had been parameterized with the parameters a_1, a_2, a_3 and a_4.

The resulting function for the model can be written as:

$$CEF_m(n_{\mathrm{bs}}) = a_1 + (a_2 - a_1) \cdot \frac{1}{2} \left[1 + \mathrm{erf}\left(a_3 \cdot \left[\mathrm{ld}\left(n_{\mathrm{bs}} \right) - a_4 \right] \right) \right] \qquad (4.4)$$

with

$$\mathrm{erf}(x) = \frac{2}{\sqrt{\pi}} \cdot \int_0^x e^{-\tau^2} d\tau \qquad (4.5)$$

The parameters have a distinct meaning for the modeling, as indicated in Figure 4.11. With the knowledge gained in Section 4.1.1, some predictions on the influence factors of the parameters can also be done. a_1 is the value the function aspires for $n_{\mathrm{bs}} \to -\infty$. This should correspond to the influence of the DRR. For high block sizes $n_{\mathrm{bs}} \to \infty$, the function reaches the value a_2, which should represent the SNR. a_3 represents the steepness of the function in the range of the turning point. The position of the turning point is represented by a_4. The last two parameters seem to depend on the reverberation time T.

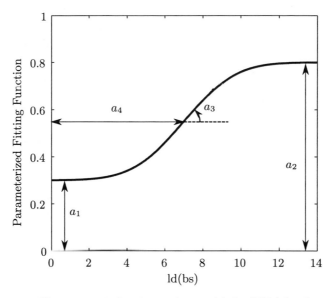

Figure 4.11: The parametric function used to model the CEF (after SCHMOCH (2011))

Determination of the parameters

The database of simulated room impulse responses and corresponding CEF functions already used for the training and evaluation of the neural network was used to determine the connection between the four parameters $a_{1..4}$ and SNR, DRR and T. For this purpose, for every simulated CEF, a curve-fitting with the modeled CEF_m after (4.4) was performed. Every curve-fitting resulted in a set of estimated parameters for the simulated SNR, DRR and T. The curve fitting was realized by an iterative optimization. The optimization criterion was the least-squared error so that

$$\min \|CEF - CEF_m(a_{1..4})\| \to a_{1..4} \qquad (4.6)$$

An extract of the results is shown in Figure 4.12. The dependencies of the parameters are similar to the predictions made in Figure 4.7 but also indicate some cross dependencies. Further details can be found in (SCHMOCH, 2011).

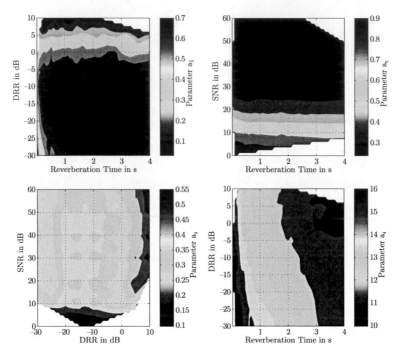

Figure 4.12: The major dependencies between a_i and SNR, DRR and T (SCHMOCH, 2011)

The final aim is the estimation of T. It is necessary to model at least one of the dependencies between the parameters a_i and the reverberation time. This model would hopefully allow an inversion and finally a calculation of T from the fitted CEF_m. The most obvious influence of T can be found in the parameter a_4. Therefore, the dependency between a_4 and T has been modeled in several approaches. As can be seen in Figure 4.12, there is also a slight dependency of a_4 on DRR, which also had to be included in the model to get reasonable results. The final, best solution found is

$$a_4(T, \text{DRR}) = 6.58 \cdot (T/\text{s})^{0.24} - 2.056 + 5.611 \cdot (\text{DRR}_{\log}/\text{dB} + 46)^{0.13} \qquad (4.7)$$

Due to the dependency on T and DRR at the same time, either DRR has to be known to calculate T, or another dependency has to be added to gain knowledge about T and DRR on the same time. As a_1 seems to depend on DRR, it was chosen as second function and modeled as

$$a_1(T, \mathrm{DRR_{log}}) = 0.41 + 0.39 \cdot \mathrm{erf}(0.12(\mathrm{DRR_{log}}/\mathrm{dB} - 4.81))$$
$$+ e^{-(3.1(T/s + 0.76) + 0.05(\mathrm{DRR_{log}}/\mathrm{dB} + 0.9))} \tag{4.8}$$

$$\mathrm{DRR_{log}} = 10 \log(\mathrm{DRR}) \tag{4.9}$$

The equation system formed by these approximations cannot be solved analytically for T and DRR. For the calculation of T and DRR, a numeric method for solving nonlinear equation systems using a Levenberg-Marquardt algorithm was used.

4.2.4 Analytical Approach

For the estimation of the spectral densities, the time signal is cut into segments, which have the duration t_{bs}. Every segment is multiplied with a time window function $w(t)$, often a *Hann* window. The averaging in the time domain of blocks with the duration t_{bs} can be approximated as a convolution of the complex spectrum with the transformation of the window function in the frequency domain together with a multiplication with a Dirac comb. The multiplication represents nothing else than a limited frequency resolution. A rough approximation for the window function would be a rectangular function of the width f_{psd}. Figure 4.13 shows the time and frequency representations of a time signal, the window function and the Dirac comb.

The cross spectral density estimate between two signals $x(t)$ and $y(t)$ that result from one source signal $g(t)$ and are transmitted over two different transfer paths $h_x(t)$ and $h_y(t)$ can be expressed as

$$S_{xy} = \mathcal{F}\left\{ \frac{1}{N} \sum_{n=1}^{N} (x(-t) * y(t))) \cdot (w(t) * \delta(t - n \cdot t_{bs})) \right\} \tag{4.10}$$

$$= \mathcal{F}\left\{ \frac{1}{N} \sum_{n=1}^{N} g(t) * (h_x(-t)^* * h_y(t))) \cdot (w(t) * \delta(t - n \cdot t_{bs})) \right\} \tag{4.11}$$

where N is the number of blocks evaluated.

With the assumption that the systems h_x and h_y can be treated as linear and

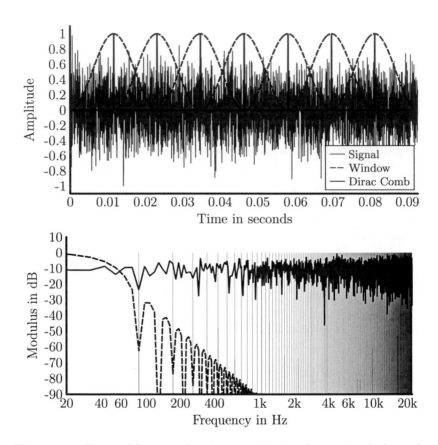

Figure 4.13: Time and frequency domain representations of a time signal, the window function and the dirac comb used in the PSD estimation.

time invariant within t_{bs}, this can be written in the frequency domain as:

$$S_{xy} = \frac{1}{N} \sum_{n=1}^{N} G(f) \cdot H_{x,n}(f)^* \cdot H_{y,n}(f) * W(f) \cdot \mathrm{e}^{(2\pi f \cdot n \cdot t_{\mathrm{bs}})} \tag{4.12}$$

$$= \mathrm{III}(f \cdot t_{\mathrm{bs}}) \cdot G(f) \cdot \frac{1}{N} \sum_{n=1}^{N} (H_{x,n}(f) \cdot H_{y,n}(f)) * W(f) \tag{4.13}$$

Using this equation in the coherence calculation (2.53), the excitation signal G cancels out (assuming it is finite and not zero) so that the coherence depends merely on the transfer paths.

$$\gamma_{xy} = \mathrm{III}(ft_{\mathrm{bs}}) \frac{\frac{1}{N} \sum_{n=1}^{N} (H_{x,n}^* H_{y,n}) * W(f)}{\sqrt{\left(\frac{1}{N} \sum_{n=1}^{N} (H_{x,n} H_{x,n}^*) * W(f)\right)\left(\frac{1}{N} \sum_{n=1}^{N} (H_{y,n} H_{y,n}^*) * W(f)\right)}} \tag{4.14}$$

In case of a time invariant system, the time averaging is no longer necessary and the signal S cancels down from this formula. The coherence can also be expressed using only the transfer functions.

$$\gamma_{xy} = \mathrm{III}(ft_{\mathrm{bs}}) \frac{(H_x^* H_y) * W(f)}{\sqrt{((H_x H_x^*) * W(f))((H_y H_y^*) * W(f))}} \tag{4.15}$$

Time domain model

A model of the CEF can be generated in the time domain based on (4.15). The impulse responses h_x and h_y can be rewritten as an addition of an early and a late part:

$$h_x(t) = h_{x,\mathrm{early}}(t) + h_{x,\mathrm{late}}(t) \tag{4.16}$$

$$h_y(t) = h_{y,\mathrm{early}}(t) + h_{y,\mathrm{late}}(t) \tag{4.17}$$

the time chosen for the separation of the two parts is chosen as t_{bs}, so that

$$h_{\mathrm{early}}(t) = 0 \text{ for } t > t_{\mathrm{bs}} \tag{4.18}$$

$$h_{\mathrm{late}}(t) = 0 \text{ for } t \leq t_{\mathrm{bs}} \tag{4.19}$$

According to SCHROEDER (1996) these parts are not correlated in a room impulse response. With the assumption that the energy distribution for both sensors is similar, the coherence of the total signal can then be written as an addition of the coherences from the two parts using (2.77).

$$\gamma_{xy} = \gamma_{xy,\text{early}} \cdot \left(\frac{E_{\text{early}}}{E_{\text{early}} + E_{\text{late}}} \right) + \gamma_{xy,\text{late}} \cdot \left(\frac{E_{\text{late}}}{E_{\text{early}} + E_{\text{late}}} \right) \quad (4.20)$$

$$= \gamma_{xy,\text{early}} \cdot \left(\frac{1}{1 + \frac{E_{\text{late}}}{E_{\text{early}}}} \right) + \gamma_{xy,\text{late}} \cdot \left(\frac{1}{1 + \frac{E_{\text{early}}}{E_{\text{late}}}} \right) \quad (4.21)$$

The whole energy of all reflections arriving at time t in an impulse response with a reverberation time T can be assumed as (KUTTRUFF, 2000)

$$E(t) = E_0 \cdot e^{-6 \cdot \ln 10 \cdot \frac{t}{T}} \quad (4.22)$$

As the absolute energy is not important, the energy of the reverberant part of the impulse response is normalized, so that

$$E_{\text{reverb}} = \int_0^\infty E(t')\mathrm{d}t' = E_0 \frac{T}{6 \ln 10} = 1 \quad (4.23)$$

$$\to E_0 = \frac{6 \ln 10}{T} \quad (4.24)$$

To get the total energy in the first time t of the impulse response, an integration is necessary.

$$E_{\text{earlyreflections}}(t) = \int_0^t E(t')\mathrm{d}t' \quad (4.25)$$

$$= \left(1 - e^{-6 \cdot \ln 10 \cdot \frac{t}{T}} \right) \quad (4.26)$$

This does not include the direct sound. The relation of the direct sound to the total energy in the impulse response is determined by the DRR (2.29). Accordingly, the normalized energy of the direct sound is:

$$E_{\text{direct}} = \text{DRR} \cdot E_{\text{reverb}} = \text{DRR} \quad (4.27)$$

In combination with the energy of the early reflections, the total normalized energy in the early part of the impulse response is:

$$E_{\text{early}}(t) = E_{\text{earlyreflections}} + E_{\text{direct}}(t) \tag{4.28}$$

$$= \left(1 - e^{-6 \cdot \ln 10 \cdot \frac{t}{T}}\right) + \text{DRR} \tag{4.29}$$

The energy of the second part is accordingly:

$$E_{\text{late}}(t) = \int_{t}^{\infty} E(t')\mathrm{d}t' \tag{4.30}$$

$$= e^{-6 \cdot \ln 10 \cdot \frac{t}{T}} \tag{4.31}$$

and

$$E_{\text{early}} + E_{\text{late}} = 1 + \text{DRR}; \tag{4.32}$$

The relation of the energies can be expressed as:

$$\frac{E_{\text{early}}(t)}{E_{\text{late}}(t)} = \frac{1 - e^{-6 \cdot \ln 10 \cdot \frac{t}{T}} + \text{DRR}}{e^{-6 \cdot \ln 10 \cdot \frac{t}{T}}} \tag{4.33}$$

$$= (1 + \text{DRR}) \cdot e^{6 \cdot \ln 10 \cdot \frac{t}{T}} - 1 \tag{4.34}$$

With this energy distribution, the coherence can be expressed as:

$$\gamma_{xy} = \gamma_{xy,\text{early}} \cdot \left(\frac{1}{1 + \frac{1}{(1+\text{DRR}) \cdot e^{6 \cdot \ln 10 \cdot \frac{t}{T}} - 1}}\right) + \gamma_{xy,\text{late}} \cdot \left(\frac{1}{(1 + \text{DRR}) \cdot e^{6 \cdot \ln 10 \cdot \frac{t}{T}}}\right) \tag{4.35}$$

Further, from now t is chosen as $t = t_{\text{bs}}$ so that it agrees with the block size used for the PSD calculation. This definition allows us an estimation of $\gamma_{xy,\text{early}}$ from (4.15). The impulse response h_{early} is cut after t_{bs}. The corresponding frequency resolution is $\Delta f = \frac{1}{t_{\text{bs}}} = f_{psd}$. This means that the window function W used for the convolution is exactly as wide as the frequency resolution of the transfer function. The convolution can be ignored and the transfer functions cancel down to unity. This finally leads to an estimate for the first part of the impulse response.

$$\gamma_{xy,\text{early}} = 1 \tag{4.36}$$

A prediction of $\gamma_{xy,\text{late}}$ is not that obvious. The coherence can be assumed as that in a perfectly diffuse sound field by (2.58) or (2.59), although this is only met partially in real rooms and only for certain placements of the sensors. An basic estimate for $\gamma_{xy,\text{late}}$ therefore is:

$$\gamma_{xy,\text{late}} = 0 \tag{4.37}$$

This finally leads to an estimate of the CEF as a function of the PSD block size t_{bs}

$$CEF = 1 \cdot \left(\frac{1}{1 + \frac{1}{(1+\text{DRR}) \cdot e^{6 \cdot \ln 10 \cdot \frac{t_{\text{bs}}}{T}} - 1}} \right) \tag{4.38}$$

Equation (2.77) can be used to also include the influence of noise as expressed by the SNR

$$CEF = \frac{1}{1 + \frac{1}{(1+\text{DRR}) \cdot e^{6 \ln 10 t_{\text{bs}}/T} - 1} + \frac{1}{\text{SNR}} \cdot (1 + \frac{1}{(1+\text{DRR}) \cdot e^{6 \ln 10 t_{\text{bs}}/T} - 1})} \tag{4.39}$$

Frequency domain model

Alternatively, the CEF can be modeled in the frequency domain. Equation (4.15) shows that the coherence can be calculated from the transfer functions by a convolution with the transformation of the window function. This is similar to a moving average. The *Hanning* function in the frequency domain is displayed in Figure 4.13. This function is now approximated by a *rect* function of the width $\frac{1}{t_{\text{bs}}}$.

The coherence estimate for one frequency can then be expressed as

$$\gamma_{xy}(f) = \frac{\frac{1}{N_b} \sum_{f_l}^{f_u} S_{xy}}{\sqrt{\frac{1}{N_b} \sum_{f_l}^{f_u} S_{xx} \cdot \frac{1}{N_b} \sum_{f_l}^{f_u} S_{yy}}} \tag{4.40}$$

where N_b is the number of frequency bins in the range between f_l and f_u

$$f_l = f - \frac{1}{2t_{bs}} \tag{4.41}$$

$$f_u = f + \frac{1}{2t_{bs}} \tag{4.42}$$

$$N_b = \frac{t_{bs}}{t_{ir}} \tag{4.43}$$

t_{ir} is the duration of the impulse response, which is proportional to T so that $t_{ir} = a \cdot T$. The number of modes N_m in the same frequency range depends on the room volume V

$$N_m = \frac{4V \left(f_u^3 - f_l^3\right)}{3c^3} \tag{4.44}$$

That means, the range between f_l and f_u contains N_b bins and N_m modes. If $N_m > N_b$, the frequency bins are uncorrelated to each other. This is valid in a diffuse sound field, which is partly valid in most rooms above the critical frequency (2.2).

The normalization in (4.15) is not affected by the running average. For frequencies above f_c the problem therefore can be reduced to the mean value over a number of uncorrelated complex numbers with a magnitude of one.

The expectation mean value over n complex numbers is

$$\mathrm{E}\left(\frac{1}{n}\sum_{i=1}^{n} c_i\right) = \frac{1}{\sqrt{N}} \quad \text{with} \quad c_i \in \mathbb{C} \quad \text{and} \quad ||c_i|| = 1 \tag{4.45}$$

This does not include the direct sound, which has to be considered according to the DRR. This leads to an approximation of the CEF

$$CEF = \frac{1}{(1+\mathrm{DRR})} \cdot \left(\mathrm{DRR} + \sqrt{\frac{t_{bs}}{t_{ir}}}\right) \tag{4.46}$$

Sensor noise also has to be considered according to the SNR:

$$CEF = \frac{1}{(1+\mathrm{DRR}) \cdot (1 + \frac{1+\frac{1}{\mathrm{DRR}}}{\mathrm{SNR}})} \cdot \left(\mathrm{DRR} + \sqrt{\frac{t_{bs}}{t_{ir}}}\right) \tag{4.47}$$

Comparison

A comparison of the modeled CEFs with measured ones is used to determine the similarities and deviations. A first comparison did not show a high agreement and indicated that the theoretic approach does not fully cover all influences in the CEF. The introduction of empirically determined adjustment factors however increases the agreement significantly. (4.39) was changed to

$$CEF_{\text{time}} = \cfrac{1}{1 + \cfrac{1}{(1+\text{DRR})\cdot e^{0.7(6\ln 10 t_{\text{bs}}/T)^{1/2}} - 1} + \frac{1}{\text{SNR}} \cdot \left(1 + \cfrac{1}{(1+\text{DRR})\cdot e^{0.7(6\ln 10 t_{\text{bs}}/T)^{1/2}} - 1}\right)}$$

$$(4.48)$$

and (4.47) was adapted to fit the measured CEFs by introducing adjustment factors as

$$CEF_{\text{freq}} = \frac{1}{(1 + \text{DRR}) \cdot \left(1 + \frac{1 + \frac{1}{\text{DRR}}}{\text{SNR}}\right)} \cdot \left(\text{DRR} + \sqrt{\frac{2t_{\text{bs}}}{t_{\text{ir}}}}\right) \qquad (4.49)$$

Figure 4.14 shows a comparison of the models according to (4.48) and (4.49), along with the PU- and PP-CEF. In all three situations the SNR was set to 20 dB and the DRR to -20 dB. The sensor setup was that of the hearing air dummy head described in Section 2.2.3. The simulated impulse responses were convolved with a noise signal of the duration 2^{23} samples (190.2 s), which is long enough to allow averaging even for the big block sizes in the CEF calculation. The sound velocity for the PU-CEF was estimated from the two pressure signals from one hearing aid. For the PU-CEF the frequency range between 100 Hz and 1 kHz was evaluated, as for higher frequencies the sound velocity estimation introduces some errors. For the PP-CEF the signals of the front microphone from both BTE hearing aids were analyzed. The frequency range between 400 Hz and 20 kHz was evaluated, as for lower frequencies the spatial coherence of the binaural signal in a diffuse sound field does not decrease as necessary for the reverberation time estimation.

Figure 4.15 shows the influence of SNR and DRR changes on the model. The reverberation time was set to 3 s for all three examples. The SNR is varied between 20 dB and 0 dB. The DRR is varied in the range between -20 dB and 0 dB.

The comparison shows that both models do not fit perfectly. For a high SNR and

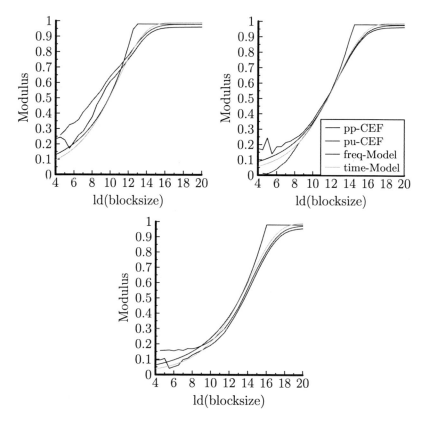

Figure 4.14: Comparison of measured and modeled CEFs for three different reverberation times. Top left: 0.3 s, Top right: 1 s, Bottom: 3 s

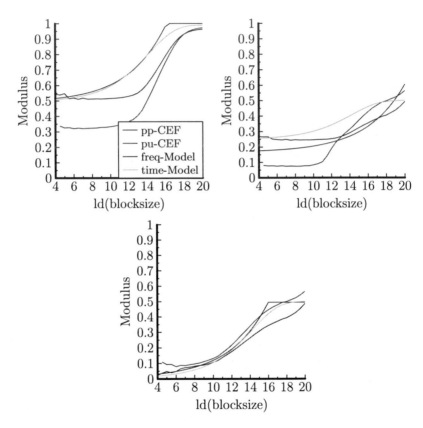

Figure 4.15: Comparisson of measured and modeled CEFs for three different SNR and DRR combinations. Top Left: SNR = 20 dB, DRR = 0 dB, Top right: SNR = 0 dB, DRR = 0 dB, Bottom: SNR = 0 dB, DRR = −20 dB.

a low DRR they match quite good, but for the other cases there are significant deviations. Anyhow, the trends of the models are both correct and there is a good agreement with the measured CEF for both models in most situations. Also, the PU- and PP-CEF do not match each other. This most possibly results from the estimation of the sound velocity from the sound pressures.

As the time domain based model seems to be a better fit than the frequency domain model, the time domain model is used for all further evaluations.

Estimation

The method for the estimation of the reverberation time from a CEF is the same, no matter whether a PP- or a PU-CEF is used and independent of which of both analytical CEF models is used. The reverberation time estimation is based on a curve fitting by minimizing the mean squared error between the measured and the modeled CEF so that

$$\min \|CEF - CEF_m(T, \text{SNR}, \text{DRR})\| \to T, \text{SNR}, \text{DRR} \qquad (4.50)$$

Another approach, which is not further evaluated here, would be to estimate the SNR from the coherence estimate as described by JEUB (2012) and the DRR as described by KUSTER (2011). Both values can also directly be estimated from the CEF or the sound field classification described in Chapter 3. For low block sizes the CEF is dominated by the DRR and for high block sizes it is dominated by the SNR. With this a priori knowledge, (4.39) can be solved for T and the reverberation time can be calculated directly from the CEF.

4.3 Considerations on Real Time Implementations

For a real time estimation of the reverberation time in a small microphone array there are some important considerations on the calculations. First, the SDs of the input signals necessary for the coherence estimation need to be estimated live. Due to memory constrains, usually an exponential average as described in 2.5.1 is used. This approach has also the advantage that it deals well with a limited amount of time variances. To increase the dynamic of the spectral density estimation, usually an overlap is used. That means the analysis blocks overlap to a certain amount α_{ol}. The higher the overlap, the better the dynamic of the spectral density estimation will be. An increase in overlap will also lead to an increase in computational costs, as more blocks have to be calculated in the same time. Figure 4.16 shows the increase in computational cost as a function of overlap according to SCHMOCH (2011). Obviously, a very high overlap is very costly and therefore not advisable. In addition, with a high overlap the differences between two blocks become insignificant, which also reduces the benefits of the overlap.

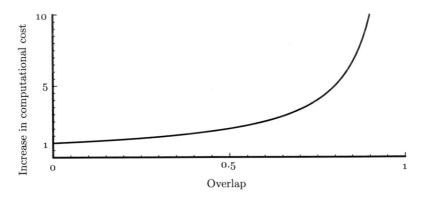

Figure 4.16: Increase in computational cost of the SD estimation due to overlap of blocks

For the calculation of the CEF not only one spectral density estimation with one block size has to be performed, but the estimation of two PSDs and the cross SD between the two signals. And this calculation has to be performed for each evaluated block size. Therefore, a multitude of spectral density estimations has to be performed in real time. Depending on the layout, the calculation of multi spectral density has to be performed in on time step. With an optimal arrangement of the calculations, the computational load can be stretched out over time, so that the load stays more constant over time (SCHMOCH, 2011). One example of an aligned and an optimal calculation arrangement is shown in Figure 4.17.

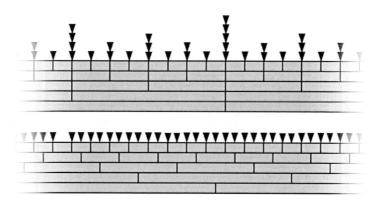

Figure 4.17: Optimal distribution of calculation of the SD blocks over time. Every arrow indicates an SD calculation. Top: Aligned calculation; Bottom: Optimal arrangement (SCHMOCH, 2011)

4.4 Error Detection and Post-Processing

The estimation of the reverberation time can be significantly improved by equipping the algorithm with an error detection. Typical errors – i.e., those for which one can implement a recognition – occur, for example, in a low SNR or too large DRR and unstable CEFs.

In the first step, a present CEF can be adjusted. Values that are based on a too brief averaging can be ignored or considered in the following approximation to a lesser degree. Special situations also arise where the CEF does not increase monotonically. Areas with such an increase can also be ignored or weighted less.

In a second step the CEF already indicates whether a reverberation time estimation seems promising or not. If the CEF for high block sizes is too low, for example below 0.2, this indicates a very low SNR and the estimation will most probably fail or return strange results. The same goes for a CEF with very high values for low block sizes. In this case the DRR is most probably very high so that the signal does not include enough reverberation for a valid estimation. In addition, similar CEF values for low and high block sizes indicate a combination of a high DRR and a low SNR. In that case the estimation usually will not return valid results. Most of these errors can be explained by the influences of SNR and

DRR on the CEF as indicated by Figure 4.7.

In a next step, the actual estimates are checked for plausibility. Limits for these values limit the estimate, but they also increase the reliability of the algorithm. If an error is detected, the estimation is canceled and the estimated value is discarded or marked as unreliable.

For a real time estimation the results are smoothed by averaging over a period of the last 3 s. This is plausible, as the reverberation time in one room is usually quite constant, although it may also show some slight deviations, as do the measured results depending on measurement setup and evaluation software (BORK, 2005; KATZ, 2004). The smoothing reduces the adaption rate in case of a change, e.g. a room change. Discharged or unreliable estimation values are not considered in the averaging. The standard deviation in this period can be used to detect sudden changes, like a room change or the closing or opening of doors. SCHMOCH (2011) also showed that by substracting the standard deviation from the estimated value the estimation performance can be further increased.

4.5 Verification and Examples

4.5.1 Static Scenes

For the verification of the blind reverberation time estimation, the estimation results are compared to measured results for rooms with measured impulse response. The binaural impulse responses utilized for this task were taken from the Aachen impulse response database (AIR) (JEUB, SCHÄFER, and VARY, 2009). The impulse responses in the AIR database were recorded with a different dummy head than the one used for the training of the neural network and the fitting of the empirical model (see Section 2.2.3). All binaural room impulse responses from the AIR database were evaluated. For every room and every source-receiver combination, the early decay time (EDT) as well as T_{10} to T_{60} have been calculated in octave bands according to ISO-3382-2 (2008) using the methods of the *ITA-Toolbox*. Not all reverberation times could be calculated this way for all situations as for some the SNR was too bad. The single values used for comparison are calculated as the median over all valid reverberation times in all bands.

In addition, for each combination the reverberation time has been estimated using the three methods previously described. Each impulse response was convolved with a sequence of white noise with a duration of 95.1 s. No additional noise was added to the single sensor signals. The binaural room impulse responses are not suitable for a valid calculation of the sound velocity, therefore the PP coherence was evaluated. Due to the sensor or ear distance, frequencies below 400 Hz could not be evaluated. The time constant for the spectral density estimation t_c was set to 5 s, which leads to a rather slow adaption but stable and reliable results. The first 10 s of every evaluation have been dismissed, as the CEF is not stable in this range. The CEF was calculated in octave bands. The resulting CEF has been used by all three estimation methods. All methods used the post-processing and error detection methods described in Section 4.4. Unreliable results were marked and not used for further evaluation.

For the estimation results, the median value over all reliable estimates in the frequency range between 400 Hz and 10 kHz has been calculated. Figure 4.18 shows the results of all four methods together with the calculated EDTs for all rooms in the AIR database. The EDT shows the highest correlation of all evaluated metrics as explained in Section 4.5.2.

Figure 4.19 shows a comparison of the reverberation times estimated using the different estimation layouts, and those calculated using ISO-3382-2 (2008) separated by room and estimation method. There is a concordance between the estimation of the reverberation time and the measured EDT for most rooms. The estimation methods deliver similar results, which indicates that the deviations result more from the CEF than the estimator itself. The spreading in the estimated reverberation times of one room is also higher than in those calculated using ISO-3382-2 (2008).

The span of results for measured and estimated results is indicated in Figure 4.20. In the *Aula Carolina* the measured and estimated reverberation times are very similar. The spreading of the EDT in dependence of the source to receiver distance is also similar. In the *Booth* the measured reverberation time is very short, almost non existent. All estimation methods deliver a similar, higher estimate. In the *Lecture Hall* the EDT is shorter than the estimation methods predict. Besides, it is very constant over the different source and receiver positions, whereas the estimation methods show quite some variation over the different situations. The estimated reverberation times in the *Meeting Room* and *Office Room* are both slightly overestimated by all three methods. In all cases but the *Stairway* all three estimation methods return similar results and similar spreadings of estimation

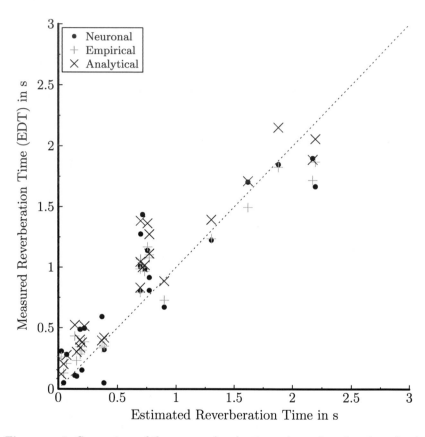

Figure 4.18: Comparison of the measured and estimated reverberation times for six rooms and different combinations of source and receiver positions.

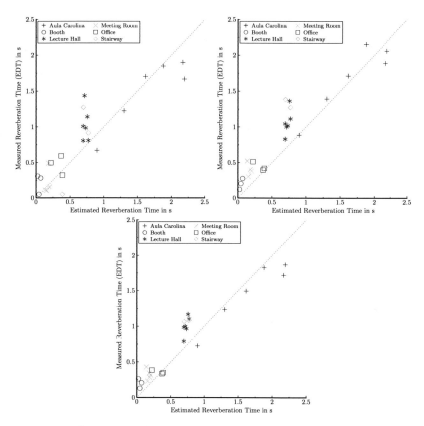

Figure 4.19: Comparison of the measured and estimated reverberation times for six rooms and different combinations of source and receiver positions. Top left: Neuronal, Top right: Empirical, Bottom: Analytical

results. In case of the *Stairway* all estimation methods return quite different results, with the *Neuronal Network* being closest to the measured estimation time, though with a spreading of estimated results of over 1 s.

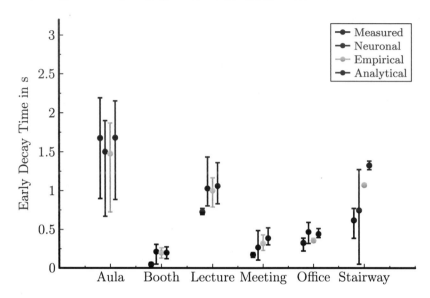

Figure 4.20: Comparison of the measured and estimated reverberation times for six rooms and different combinations of source and receiver positions. For each room the span of reverberation times for the different combinations is indicated with the mean value indicated by the dot.

4.5.2 Error Sources

SCHMOCH (2011) performed an extensive evaluation of influence factors on the reverberation time estimation. There are basically two kinds of errors that influence a reverberation time estimation based on a CEF. The first kind of error distorts the CEF itself. Accordingly, independent of the estimation method the result will not be reliable. The second kind of error results from a faulty estimator. The second error occurs, for example, for a neural network estimator with faulty or fragmentary training data.

There is a group of influences that render any reliable estimation of the reverberation time impossible. A too low SNR or too high DRR are examples for these influences. Such errors can be detected as already discussed, but they cannot be avoided or compensated.

Testing the estimation algorithm on the same data already used for the learning of the algorithm already leads to some deviations between measured and estimated reverberation times. This leads to the assumption that there are some factors that are not considered in the simulation of the training data but influence the result of the estimation. One possible example are the details in the impulse responses, like the directions and delays of the early reflections. The deviation further increases when examining real audio examples instead of simulated ones. This leads to the assumption that further influential factors are not considered in the simulations that influence the performance of real situations. Possible examples could be modes or non-exponentially decaying reverberation slopes. Finally, the excitation signal also influences the estimation result, pauses in the signal or narrow-band excitations lead to errors in the estimation. SCHMOCH (2011) also found that strong early reflections influence the estimation results, whereas the direction of the direct sound has no direct influence on the estimation.

Figure 4.21 shows the mean, minimum and maximum absolute deviations between the between the estimated reverberation times and the measured EDT after ISO-3382-2 (2008). For this purpose, the differences between the estimation and measurement in each situation and room were calculated. Obviously there are some significant deviations with a typical estimation error in the range of 0.2 s. Different estimators deliver the best performance, depending on the room. In total the empirical model seems to deliver the most reliable results.

Figure 4.22 shows the correlation coefficients between the estimated reverberation times and the different metrics of measured reverberation times after ISO-3382-2 (2008). For this purpose, the correlation between the results for all rooms and all situations from the estimation processes and the measurements was calculated. The correlation between the estimated times and the short time metrics is higher than the one for the longer time metrics. The EDT shows the highest correlation with all estimation processes. With increasing evaluation time the correlation decreases. This also emphasizes the strong influence of the early reflections in the impulse response on the estimation process. Accordingly, the CEF is mostly influenced by the early part of the impulse response, which includes the direct sound and the early reflections.

Basically, Figure 4.22 indicates that the reverberation time estimation estimates the EDT. It also indicates, that most rooms in the AIR database have impulse

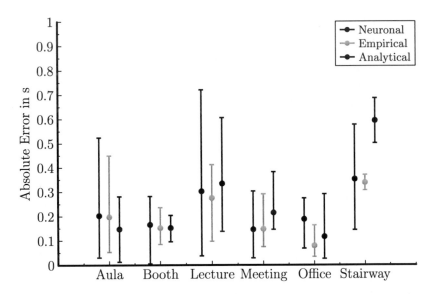

Figure 4.21: Minimal, mean and maximal absolute error in the comparison between the estimated reverberation times and the EDT.

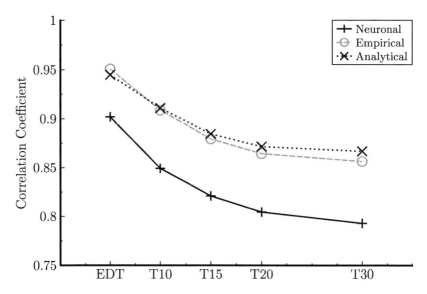

Figure 4.22: Correlation coefficient between the estimated reverberation times and the measured reverberation times after different metrics for all evaluated situations

responses with a non-linear energy decay. Otherwise there would not be such big deviations between the EDT and T_{20}, for example. Figure 4.23 shows the mean reverberation times after different metrics for the rooms in the AIR database. The *Aula Carolina* and the *Stairway* show a strong dependency on the evaluated part of the energy decay curve. In these rooms especially the EDT shows a very broad spectrum of values, depending on source and receiver positions. With increasing evaluation time the spreading of the values decreases as the evaluated reverberation time increases. GADE (1994) showed that most auditoria do not have a really diffuse sound field. In addition, rooms that are very long or coupled volumes do not fulfill the criteria defined by SABINE (1922) on the exponential sound energy decay. The different reverberation time metrics differ for all rooms with an non-exponential energy decay (KUTTRUFF, 2000; XIANG et al., 2011). Moreover, the EDT is strongly influenced by the direct sound and very early reflections, and accordingly strongly depends on the source to receiver distance and positions. In conclusion, the results from Figure 4.22 indicate that none of the rooms shows a perfectly exponential sound energy decay, which is also to be expected for most real rooms.

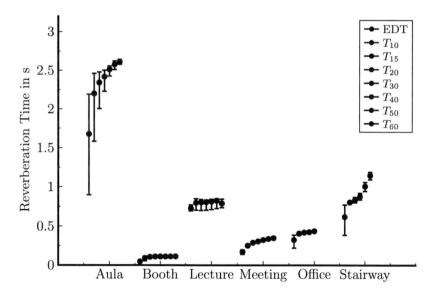

Figure 4.23: Measured average reverberation times of the AIR database after different metrics

4.5.3 Variant Scenes

The estimation method is able to do a live real-time estimation each time the CEF is updated. Figure 4.24 shows the results of such a live estimation using the empirical estimator in a sequence with four room changes. The audio sequence was generated by appending four measured sound sequences from different rooms. In each room, the impulse response was measured for the calculation of a reference reverberation time. The empirical approach was used to estimate the reverberation time every 0.1 s. Unreliable results where compensated by a simple sample-and-hold. Each scene was time invariant, no movement of source or receiver took place. In contrast to the previous evaluations that used white noise as excitation signal, the signal used in this case was a speech sequence. The time constant t_c for the spectral density estimation was set to 1 s, which also leads to some smoothing of the results. The calculation was performed in octave bands. The single-value result is the median from the result in the frequency range between 400 Hz and 10 kHz.

The results in Figure 4.24 show that there are some significant up- and downturns over time in every situation. For every room change, the method also needs some seconds to adapt to the new room. Some deviations from the measured reverberation time are significant. The strong break-in at 33 s occurs due to a pause in the speech signal.

4.6 Discussion

A new method for the estimation of the reverberation time based on the coherence estimate function (CEF) was introduced. The CEF describes the influence of the block size used for the coherence estimation on the result of the estimation. For the reverberation time estimation, the CEF between two distributed pressure sensors or a pressure and a velocity sensor can be utilized. It could be shown that the CEF depends strongly on the room acoustic properties T, DRR and SNR.

Two different models were developed for the CEF based on room acoustic properties, but both do not fully describe the behavior of the CEF.

A method for the calculation of the CEF in real time from live input signals was described. Based on the CEF, three different estimators were developed and

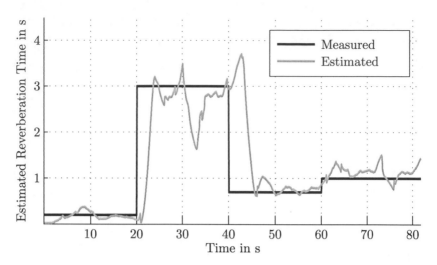

Figure 4.24: Live estimation of the reverberation time in an audio sequence with four abrupt room changes.

their performance compared. One estimator utilized a *neural network* that was trained by the results of a Monte Carlo simulation of different room acoustic situations. The *emipircal* estimation uses a curve fitting of an error-function to the CEF. The parameters of the empirical model were achieved from the training data already used in the *neural network*. In a final approach, the dependency between the room acoustic properties and the CEF was modeled analytically in the time- and frequency-domain to understand the underlying dependencies.

A comparison of estimation results with the results from a reverberation time calculation using ISO-3382-2 (2008) was performed using measured binaural impulse responses from the AIR database (JEUB, SCHÄFER, and VARY, 2009). All three developed estimators deliver similar performance results, with the *neuronal network* showing the poorest performance and the *empirical* and *analytical* estimators showing very similar results. A comparison of the estimation results with different evaluation metrics showed that the estimation results show the highest correlation with the early decay time (EDT) and a decreasing correlation with reverberation times using higher evaluation ranges.

The method is able to estimate the reverberation time from live signals. A test with a time variant scene including three room changes showed that the estimator needs up to 10 s to adapt to the new scene. In addition, gaps in the excitation signal lead to errors in the estimation. The live estimation could be improved by further post-processing, error detections and a priori knowledge from other sensors.

5

Conclusion

5.1 Summary

One global objective in the signal processing in mobile devices such as hearing aids or mobile phones is the improvement of their performance in different, foremost noisy, situations. One key factor for this is the correct classification of the situation, the signal and environment and the user preferences. The aim of this work was to improve the classification and control blocks of mobile devices by gathering information on the acoustic environment. The focus lies on a local observation of the sound field around the device and a global description of the acoustic environment in form of the reverberation time.

The work started with an analysis of the sound field indicators and their behavior in different acoustic situations. Based on these observations a sound field classification was proposed. After a review of existing methods, for the description of the signal and the acoustic environment along room acoustic and signal processing literature, the four classes *free*, *diffuse*, *reactive* and *noise* were chosen as target classes for the classification. The class *noise* is not really a sound field, but merely the absence of any acoustic sources so that the signal perceived by the device is dominated by wind-, sensor-, and electronic-noise.

The classification depends on two spatial coherences, that of two nearby pressure signals and that of a pressure and a sound velocity signal. Two ways of calculating those input values were explained. One is based on a sound field microphone with four omnidirectional sensors arranged in the shape of a regular tetrahedron. This setup allows a sound field classification without any a priori knowledge about the situation and environment. Another approach with a hearing aid setup was evaluated. In this case, the magnitude of the estimated sound velocity in some sound fields depends on the source direction. Accordingly, the direction or the directions of the sound source(s) need to be known for the sound field classification.

For the classification, two different approaches were proposed. The first method was based on a fuzzy distance classification. It evaluated the distance between the target classes and the feature vector in a three-dimensional space consisting of the spatial coherences. The locations of the target classes in this feature space were derived from theoretical predictions, so that no training process was necessary. The second approach was based on the prediction of the resulting spatial coherences in a superposition of the basic sound fields. This prediction can be written as a matrix multiplication, and by an inversion the relative energies of the single sound fields can be calculated. The total sound field energies can be gained by a multiplication of the relative sound field energies with an estimation of the total energy based on the intensity.

The classification methods were verified with two different approaches. The fist simple test was conducted by usage of a sequence of simulated basic sound fields. Both classifiers were able to classify all four basic sound fields as expected. In a second approach, the distance-dependent sound field composition in a room with one sound source and reverberation was used for the verification. The theoretical sound field composition is known from room acoustic and diffuse field theory. The classification returns results that agree with the theoretical predictions for the dominant sound fields. The accuracy of the classification declines with the relative sound field energy so that the estimation of the energy of underlying sound fields shows some significant errors.

Further tests revealed a strong influence of a sensor mismatch on the sound field classification. For this purpose, an automatic sensor mismatch compensation was designed. The compensation is able to perfectly compensate a sensor mismatch after an initial training. After the training, the method adapts automatically to any further sensor drift due to aging or other influences. A set of examples of the sound field classification in realistic scenarios showed plausible results and demonstrated the possibility of real-time classifications with the developed demonstrator.

To gather more information on the acoustic environment, the influence of some acoustic properties on the coherence estimate function (CEF) was evaluated. The CEF describes the dependency between the coherence estimate and the block size used for this estimation. In the context of two acoustic sensors in a reverberant sound field, this function shows a typical shape. Experiments revealed that at least the reverberation time, the signal-to-noise energy ratio (SNR) and direct-to-reverberant energy ratio (DRR) have a significant influence on the CEF. These influences have distinct impacts on the CEF so that an estimation of those three parameters from a measured CEF seemed feasible.

For the estimation of the reverberation time, three different estimators were introduced. A neuronal network was used for a proof of concept of the method. The layout was a feed-forward network with two hidden layers, including 20 and 10 neurons. The network was trained using 4000 simulation results. The network adapted very well and after 21 training epochs no further gain was achieved by further training. A set of new simulations, not used for training, validated the general performance of the method.

Afterwards, an empirical and an analytical model of the CEF were developed. The empirical model was based on an erf function that was parameterized with four parameters. Using the database already used in the training of the neural network, the dependencies of the four parameters on the reverberation time was analyzed and modeled. The analytical model was built on considerations about the signal processing in the coherence estimation. The first results do not show a good agreement of the model with a real CEF. With some adaption of the model, a better agreement could be realized, although the result is not strictly an analytical model. The reverberation time was estimated using a curve fitting of the model to the measured CEF. Both models do not fit the CEF in all situations. Hence, there seem to be further influence factors that should be researched further.

All three methods were tested with binaural impulse responses from the Aachen impulse response database (AIR). The results indicate a general applicability of the method, although there are still some deviations between measured and estimated reverberation times. The highest accordance of the estimation results occurs with the early decay time (EDT). This indicates that the CEF is dominated by the first part of the impulse response. All binaural samples in the AIR database show a correlation factor of 0.95 between the EDT and the estimation results for the empirical and analytical model. The neural network shows a correlation factor of 0.9.

The method is able to perform live estimations from input signals. An example in a time-variant scene with three room changes indicates that the method needs up to 10 s to adapt to a new scene. There are also significant fluctuations in one room, especially during pauses in the speech signal. In applications of the method, those fluctuations should be handled by an improved post-processing.

Overall, this thesis introduced new methods for the automatic adaption of mobile devices to their environment. The described methods can be important input information for the classification and control segments of such devices to adjust the signal processing to the acoustic situation. This thesis introduced a new method for the classification of sound fields in small microphone arrays. The method is based on the sound field indicators introduced by JACOBSEN (1989). The principle of a sound field classification based on the SFIs and the two classification methods were not described in literature. The same goes for the reverberation time estimation based on the CEF, which had not been researched and published before.

5.2 Outlook

The new methods on sound field analysis in small microphone arrays described in this thesis were proven to work in laboratory conditions and a small set of examples. Further evaluations should be carried out with an extended set of scenarios. A set of real world examples as described by STREICH et al. (2010) would be very useful for this purpose. Those recording databases, however, usually have the significant downside that neither impulse responses nor further information on the sound field is available. This means, the results can be statistically analyzed, but there is no reference for either the sound field classification or the reverberation time estimation. This makes any evaluation of the performance of both methods rather complicated. There are comparisons of different reverberation time estimation methods like the one performed by GAUBITCH et al. (2012), but they usually deploy single channel recordings, which are not suitable for the new methods based on the CEF.

Although the estimation of the reverberation time from the CEF works with reasonable performance, the analytical models of the CEF do not fit the measured CEFs perfectly, which indicates that the analytical approach does not cover all influence factors. There are a number of approximations and simplifications in the approach, which could be the reason for such deviations and accordingly a more detailed analysis could yield further insight into the influence factors of the CEF.

Bibliography

E. ALEXANDRE, L. CUADRA, M. ROSA, and F. LÓPEZ-FERRERAS (Nov. 2007). "Feature Selection for Sound Classification in Hearing Aids Through Restricted Search Driven by Genetic Algorithms." In: *IEEE Transactions on Audio, Speech, and Language Processing* 15.8, pp. 2249–2256.

S. ALLEGRO, M. BÜCHLER, and S. LAUNER (2001). "Automatic sound classification inspired by auditory scene analysis." In: *Consistent and Reliable Acoustic Cues for Sound Analysis (CRAC)*. Aalborg, Denmark.

M. ARETZ (2009). "Specification of Realistic Boundary Conditions for the FE Simulation of Low Frequency Sound Fields in Recording Studios." In: *Acta Acustica United with Acustica* 95, pp. 874–882.

M. ARETZ (2012). "Combined Wave And Ray Based Room Acoustic Simulations In Small Rooms." Dissertation. RWTH Aachen University.

M. ARETZ, R. NÖTHEN, M. VORLÄNDER, and D. SCHRÖDER (2009). "Combined Broadband Impulse Responses Using FEM and Hybrid Ray-Based Methods." In: *EAA Auralization Symposium*. Helsinki.

J. BENESTY, M. M. SONDHI, and Y. HUANG (2008). *Springer Handbook of Speech Processing*. Berlin: Springer, p. 1176.

L. L. BERANEK (1993). *Acoustics*. Woodbury, New York: Acoustical Society of America, American Institute of Physics.

B. BERGLUND and T. LINDVALL (1995). *Community noise*. Center for Sensory Research Stockholm.

J. BLAUERT (2005). *Communication Acoustics*. Ed. by J. BLAUERT. Berlin Heidelberg NewYork: Springer.

P. BLOOM and G. CAIN (1982). "Evaluation of two-input speech dereverberation techniques." In: *IEEE International Conference on Acoustics, Speech, and Signal Processing, ICASSP*, pp. 164–167.

I. BORK (2005). "Report on the 3rd Round Robin on Room Acoustical Computer Simulation Part II: Calculations." In: *Acta Acustica united with Acustica* 91.4, pp. 753–763.

M. BÜCHLER, S. ALLEGRO, S. LAUNER, and N. DILLIER (2005). "Sound classification in hearing aids inspired by auditory scene analysis." In: *EURASIP Journal on Applied Signal Processing* 18, pp. 2991–3002.

G. C. CARTER (1987). "Coherence and Time Delay Estimation." In: *Proceedings of the IEEE* 75.2, pp. 236–255.

G. C. CARTER, C. H. KNAPP, and A. H. NUTTALL (1973). "Estimation of the Magnitude-Squared Coherence Function Via Overlapped Fast Fourier Transform Processing." In: *IEEE Transactions on Audio and Electroacoustics* AU-21.4, pp. 337–344.

T. J. COX, F. LI, and P. DARLINGTON (2001). "Extracting room reverberation time from speech using artificial neural networks." In: *Journal of the Audio Engineering Society* Vol. 49.4, pp. 219–230.

P. DIETRICH, M. GUSKI, J. KLEIN, M. MÜLLER-TRAPET, M. POLLOW, R. SCHARRER, and M. VORLÄNDER (2013). "Measurements and Room Acoustic Analysis with the ITA-Toolbox for MATLAB." In: *Fortschritte der Akustik - AIA/DAGA*. Meran, Italy.

P. DIETRICH and M. LIEVENS (2009). "How to Obtain High Quality Input Data for Auralization ?" In: *Fortschritte der Akustik - NAG/DAGA*. Rotterdam, pp. 1604–1607.

DIN-18041 (2004). *Hörsamkeit in kleinen bis mittelgrossen Räumen*. Berlin: Deutsches Institut für Normung e.V.

DIN-EN-ISO-266 (1997). *Normfrequenzen*.

DIN-EN-ISO 3382 (2006). *Messung von raumakustischen Parametern - Teil 2: Nachhallzeit in gewöhnlichen Raumen*. Berlin: Deutsches Institut fuer Normung e.V.

C. F. EYRING (1930). "Reverberation time in "dead" rooms." In: *The Journal of the Acoustical Society of America* 1.2A, pp. 217–241.

F. J. FAHY (1989). *Sound Intensity*. First. New York: Elsevier.

A. FARINA (2000). "Simultaneous measurement of impulse response and distortion with a swept-sine technique." In: *Audio Engineering Society Convention 108*. Paris, France.

W. D. FONSECA, B. S. MASIERO, S. BISTAFA, P. DIETRICH, G. QUIQUETO, L. CHAMON, and M. VORLÄNDER (2010). "Medição de uma plataforma acústica conceitual desenvolvida por diferentes instituções de pesquisa." In: *XXIII Encontro da Sociedade Brasileira de Acústica (SOBRAC).* SOBRAC. Salvador, Bahia, Brazil.

A. C. GADE (1994). "Acoustic properties of concert halls in the US and in Europe; effects of differences in size and geometry." In: *127th Meeting Acoust. Soc. Amer., Wallace Clement Sabine Centennial Symposium, Cambridge, Massachusetts, USA.* Vol. 5. 7.

N. D. GAUBITCH, M. JEUB, T. H. FALK, P. A. NAYLOR, P. VARY, and M. BROOKES (2012). "Performance Comparison of Algorithms for Blind Reverberation Time Estimation from Speech." In: *Proceedings of the International Workshop on Acoustic Signal Enhancement (IWAENC).* Aachen, Germany.

P.-A. GAUTHIER, É. CHAMBATTE, C. CÉDRIC, P. YANN, and B. ALAIN (2011). "Beamforming Regularization, Scaling Matrices and Inverse Problems for Sound Field Extrapolation and Characterization: Part I - Theory." In: *Audio Engineering Society Convention 131.* New York.

V. HAMACHER, J. CHALUPPER, J. EGGERS, E. FISCHER, U. KORNAGEL, H. PUDER, and U. RASS (2005). "Signal Processing in High-End Hearing Aids: State of the Art, Challenges, and Future Trends." In: *EURASIP Journal on Applied Signal Processing* 18, pp. 2915–2929.

P. D. HATZIANTONIOU and J. N. MOURJOPOULOS (2004). "Errors in Real-Time Room Acoustics Dereverberation." In: *Journal of the Audio Engineering Society* 52.9, pp. 883–899.

F. HEESE, M. SCHAFER, P. VARY, E. HADAD, S. M. GOLAN, and S. GANNOT (2012). "Comparison of supervised and semi-supervised beamformers using real audio recordings." In: *Electrical & Electronics Engineers in Israel (IEEEI), 2012 IEEE 27th Convention of.* IEEE, pp. 1–5.

ISO-3382-1 (2009). *Acoustics - Measurement of room acoustic parameters - Part 1: Performance spaces.* International Organization for Standardization.

ISO-3382-2 (2008). *Acoustics - Measurement of room acoustic parameters - Part 2: Reverberation time in ordinary rooms.* International Organization for Standardization, p. 17.

F. JACOBSEN (1989). "Active and reactive, coherent and incoherent sound fields." In: *Journal of Sound and Vibration* 130.3, pp. 493–507.

F. JACOBSEN (1991). "A Simple and Effective Correction for Phase Mismatch in Intensity Probes." In: *Applied Acoustics* 33.3, pp. 165–180.

F. JACOBSEN and H.-E. de BREE (2005). "A comparison of two different sound intensity measurement principles." In: *The Journal of the Acoustical Society of America* 118.3, p. 1510.

F. JACOBSEN and T. ROISIN (2000). "The coherence of reverberant sound fields." In: *The Journal of the Acoustical Society of America* Vol. 108.1, pp. 204–210.

M. JEUB (2012). "Joint Dereverberation and Noise Reduction for Binaural Hearing Aids and Mobile Phones." Dissertation. RWTH Aachen University.

M. JEUB, M. SCHÄFER, and P. VARY (2009). "A Binaural Room Impulse Response Database for the Evaluation of Dereverberation Algorithms." In: *Proceedings of International Conference on Digital Signal Processing (DSP)*. Santorini, Greece, pp. 1–4.

B. F. G. KATZ (2004). "International Round Robin on Room Acoustical Impulse Response Analysis Software 2004." In: *Acoustics Research Letters O* 5.4, pp. 158–164.

P. KENDRICK, F. F. LI, T. J. COX, Y. ZHANG, and J. A. CHAMBERS (2007). "Blind estimation of reverberation parameters for non-diffuse rooms." In: *Acta Acustica United with Acustica* Vol. 93.5, pp. 760–770.

S. KOCHKIN (2000). "MarkeTrak V : " Why my hearing aids are in the drawer ": The consumers ' perspective." In: *The Hearing Journal* 53.2, pp. 34–41.

D. KRIESEL (2007). *A Brief Introduction to Neural Networks.* available at http://www.dkriesel.com.

M. KUSTER (2008). "Reliability of estimating the room volume from a single room impulse response." In: *The Journal of the Acoustical Society of America* Vol. 124.2, pp. 982–993.

M. KUSTER (2011). "Estimating the direct-to-reverberant energy ratio from the coherence between coincident pressure and particle velocity." In: *The Journal of the Acoustical Society of America* 130.6, p. 3781.

H. KUTTRUFF (2000). *Room acoustics.* 4th. New York: Elsevier.

H. KUTTRUFF (2007). *Acoutics - An Introduction.* first edit. London and New York: Taylor & Francis.

T. Lentz, D. Schröder, M. Vorländer, and I. Assenmacher (2007). "Virtual Reality System with Integrated Sound Field Simulation and Reproduction." In: *EURASIP Journal on Advances in Signal Processing* 2007, Article ID 70540, 19 pages.

K. Levenberg (1944). "A Method for the Solution of Certain Problems in Least Squares." In: *Quart. Appl. Math.* Vol. 2, pp. 164–168.

H. W. Löllmann and P. Vary (2008). "Estimation of the reverberation time in noisy environments." In: *Proceedings of the International Workshop on Acoustic Echo and Noise Control* 4.

H. W. Löllmann and P. Vary (2009). "Low Delay Noise Reduction and Dereverberation for Hearing Aids." In: *EURASIP Journal on Advances in Signal Processing* 2009, pp. 1–10.

N. Lopez, Y. Grenier, and I. Bourmeyster (2012). "Low Variance Blind Estimation of the Reverberation Time." In: *Proceedings of the International Workshop on Acoustic Signal Enhancement (IWAENC)*, pp. 1–4.

S. L. Marple (1987). "Digital spectral analysis with applications." en. In: *Englewood Cliffs, NJ, Prentice-Hall, Inc., 1987, 512 p.* -1.

D. Marquardt (1963). "An Algorithm for Least-Squares Estimation of Nonlinear Parameters." In: *SIAM J. Appl. Math.* Vol. 11, pp. 431–441.

R. Martin, I. Witew, and M. Arana (2007). "Influence of the source orientation on the measurement of acoustic parameters." In: *Acta Acustica united Acoustica* Vol. 93, pp. 387–397.

M. J. McAuliffe, P. J. Wilding, N. A. Rickard, and G. A. O'Beirne (2012). "Effect of Speaker Age on Speech Recognition and Perceived Listening Effort in Older Adults With Hearing Loss." In: *Journal of Speech, Language, and Hearing Research* 55.3, pp. 838–847.

G. Moschioni, B. Saggin, and M. Tarabini (2008). "3-D Sound Intensity Measurements: Accuracy Enhancements With Virtual-Instrument-Based Technology." In: *IEEE Transactions on Instrumentation and Measurement* 57.9, pp. 1820–1829.

S. Müller and P. Massarani (2001). "Transfer-Function Measurement with Sweeps." In: *Journal of the Audio Engineering Society* 49.6, pp. 443–471.

P. A. Naylor and N. D. Gaubitch (2012). "Acoustic Signal Processing in Noise: It's not getting any quiter." In: *Proceedings of the International Workshop on Acoustic Signal Enhancement (IWAENC)*. Aachen, Germany.

R. Nock and F. Nielsen (Aug. 2006). "On Weighting Clustering." In: *IEEE Transactions on Pattern Analysis and Machine Intelligence* 28.8, pp. 1223–35.

P. Nordqvista and A. Leijon (2004). "An efficient robust sound classification algorithm for hearing aids." In: *The Journal of the Acoustical Society of America* 115.6, pp. 3033–3041.

A. Piersol (1978). "Use Of Coherence And Phase Data Between Two Receivers In Evaluation Of Noise Environments." In: *Journal of Sound and Vibration* 56.2, pp. 215–228.

R. Ratnam, D. L. Jones, B. C. Wheeler, W. D. O. Jr, C. R. Lansing, and A. S. Feng (2003). "Blind estimation of reverberation time." In: *The Journal of the Acoustical Society of America* Vol. 114.5, pp. 2877–2892.

W. C. Sabine (1922). "Collected Papers on Acoustics." In: *Harvard U.P., Cambridge*.

R. Scharrer and J. Fels (2012). "Fuzzy Sound Field Classification in Devices with Multiple Acoustic Sensors." In: *Proceedings of the International Workshop on Acoustic Signal Enhancement (IWAENC)*. Aachen, Germany.

R. Scharrer and J. Fels (2013). "Sound Field Classification in Small Microphone Arrays." In: *Fortschritte der Akustik - AIA/DAGA*. Meran, Italy.

R. Scharrer and M. Vorlander (2013). "Sound Field Classification in Small Microphone Arrays Using Spatial Coherences." In: *IEEE Transactions on Audio, Speech, and Language Processing* 21.9, pp. 1891–1899.

R. Scharrer and M. Vorländer (2010). "Blind Reverberation Time Estimation." In: *International Congress on Acoustics, ICA*. Sydney, Australia.

R. Scharrer and M. Vorländer (2011). "The coherence estimate function and its dependency on the room acoustic situation." In: *Fortschritte der Akustik - DAGA*. Düsseldorf, Germany, pp. 625–626.

H. E. Schmoch (2011). "Blinde Schätzung der Nachhallzeit aus der räumlichen Kohärenz." Diploma Thesis. RWTH Aachen University.

D. Schröder (2011). "Physically Based Real-Time Auralization of Interactive Virtual Environments." Dissertation. RWTH Aachen University.

M. R. SCHROEDER (1965). "New Method of Measuring Reverberation Time." In: *The Journal of the Acoustical Society of America* Vol. 37.6, pp. 1187–1188.

M. R. SCHROEDER (May 1996). "The "Schroeder frequency" revisited." In: *The Journal of the Acoustical Society of America* Vol. 99.5, pp. 3240–3241.

M. SLANEY and P. A. NAYLOR (2011). "Audio and Acoustic Signal Processing [In the Spotlight]." In: *Signal Processing Magazine, IEEE* 28.5, pp. 160–1150.

W. STEGNER (1971). *Angle of Repose.*

A. P. STREICH (2010). "Multi-label classification and clustering for acoustics and computer security." Dissertation. ETH Zürich.

A. P. STREICH, M. FEILNER, A. STIRNEMANN, and J. M. BUHMANN (2010). "Sound Field Indicators for Hearing Activity and Reverberation Time Estimation in Hearing Instruments." In: *Audio Engineering Society Convention 128.* London, UK.

O. THIERGART, G. DEL GALDO, and E. A. P. HABETS (2012). "Signal-To-Reverberant Ratio Estimation Based On The Complex Spatial Coherence Between Omnidirectional Microphones." In: *IEEE International Conference on Acoustics, Speech, and Signal Processing, ICASSP*, pp. 309–312.

A. J. VERMIGLIO, S. D. SOLI, D. J. FREED, and L. M. FISHER (2012). "The Relationship between High-Frequency Pure-Tone Hearing Loss, Hearing in Noise Test (HINT) Thresholds, and the Articulation Index." In: *Journal of the American Academy of Audiology* 23.10, pp. 779–788.

S. VESA and A. HARMA (2005). "Automatic estimation of reverberation time from binaural signals." In: *Proc. IEEE Int. Conf. Acoust., Speech, Signal Process (ICASSP).* Philadelphia (Pennsylvania), USA.

M. VORLÄNDER (2008). *Auralization: Fundamentals of Acoustics, Modelling, Simulation, Algorithms and Acoustic Virtual Reality.* First Edit. Berlin: Springer.

W. WANG, S. SANEI, and J. CHAMBERS (2005). "Penalty function-based joint diagonalization approach for convolutive blind separation of nonstationary sources." In: *IEEE Transactions on Signal Processing* Vol. 53.5, pp. 1654–1669.

P. D. WELCH (1967). "The Use of Fast Fourier Transform for the Estimation of Power Spectra: A Method Based on Time Averaging Over Short, Modified Periodograms." In: *IEEE Transactions on Audio and Electroacoustics* Vol. AU-15, pp. 70–73.

N. XIANG, P. GOGGANS, T. JASA, and P. ROBINSON (2011). "Bayesian characterization of multiple-slope sound energy decays in coupled-volume systems." In: *The Journal of the Acoustical Society of America* 129.2, pp. 741–752.

Y. ZHANG, J. A. CHAMBERS, P. KENDRICK, T. J. COX, and F. F. LI (2006). "Blind estimation of reverberation time in occupied rooms." In: *14th European Signal Processing Conference (EUSIPCO 2006)*. Florence, Italy.

Own Publications

Journal Publications (with peer-review)

M. CZAPLIK, J. KALICIAK, A. FOLLMANN, S. KIRFEL, R. SCHARRER, M. GUSKI, M. VORLÄNDER, R. ROSSAINT, and M. COBURN (2013). "Psychoacoustic analysis of noise in the intensive care unit as a multidisciplinary approach: a randomized-controlled clinical trial." In: *Critical Care* Submitted.

R. SCHARRER and M. VORLANDER (2013). "Sound Field Classification in Small Microphone Arrays Using Spatial Coherences." In: *IEEE Transactions on Audio, Speech, and Language Processing* 21.9, pp. 1891–1899.

Conference Proceedings (with peer-review)

R. SCHARRER and J. FELS (2012). "Fuzzy Sound Field Classification in Devices with Multiple Acoustic Sensors." In: *Proceedings of the International Workshop on Acoustic Signal Enhancement (IWAENC)*. Aachen, Germany.

Conference Proceedings

P. DIETRICH, M. GUSKI, J. KLEIN, M. MÜLLER-TRAPET, M. POLLOW, R. SCHARRER, and M. VORLÄNDER (2013). "Measurements and Room Acoustic Analysis with the ITA-Toolbox for MATLAB." In: *Fortschritte der Akustik - AIA/DAGA*. Meran, Italy.

P. DIETRICH, M. GUSKI, M. POLLOW, M. MÜLLER-TRAPET, B. S. MASIERO, R. SCHARRER, and M. VORLÄNDER (2012). "ITA-Toolbox – An Open Source MATLAB Toolbox for Acousticians." In: *Fortschritte der Akustik - DAGA*. Darmstadt, Germany, pp. 151–152.

P. DIETRICH, B. S. MASIERO, M. POLLOW, R. SCHARRER, and M. MÜLLER-TRAPET (2010). "MATLAB Toolbox for the Comprehension of Acoustic Measurement and Signal Processing." In: *Fortschritte der Akustik - DAGA*. Berlin, Germany, pp. 517–518.

P. DIETRICH, B. S. MASIERO, R. SCHARRER, M. MÜLLER-TRAPET, M. POLLOW, and M. VORLÄNDER (2011). "Application of the MATLAB ITA-Toolbox: Laboratory Course on Cross-talk Cancellation." In: *Fortschritte der Akustik - DAGA*. Düsseldorf, pp. 471–472.

S. FINGERHUTH, R. SCHARRER, and K. KASPER (2009). "A method for a modal measurement of electrical machines." In: *40. National Acoustic Congress, Tecniacústica*. Cádiz, Spain, pp. 1–7.

S. FINGERHUTH, R. SCHARRER, and K. KASPER (2010). "Método de análisis modal para motores eléctricos." In: *XIV COCIM. Congreso Chileno de Ingeniería Mecánica*.

M. GUSKI, R. SCHARRER, M. CZAPLIK, R. ROSSAINT, and M. VORLÄNDER (2011). "Lärmmessungen auf der Intensivstation." In: *Fortschritte der Akustik - DAGA*. Düsseldorf, Germany.

J. KALICIAK, M. CZAPLIK, A. FOLLMANN, S. KIRFEL, R. SCHARRER, M. VORLÄNDER, and R. ROSSAINT (2011). "Psychoakustische Analyse des Lärms auf der Intensivstation." In: *Deutscher Anästhesiekongress (DAC)*. Hamburg, Germany.

K. KASPER, R. DE DONCKER, R. SCHARRER, S. FINGERHUTH, and M. VORLÄNDER (2008). "Experimental modal analysis of electrical machines using electromagnetic excitation." In: *37th International Congress and Exposition on Noise Control Engineering (Internoise 2008)*. Vol. 2008. 7. Institute of Noise Control Engineering. Shanghai, China, pp. 2238–2250.

R. SCHARRER (2009). "Spatial coherence in binaural applications." In: *Fortschritte der Akustik - NAG/DAGA*. Rotterdam, Netherlands, pp. 192–195.

R. SCHARRER (2010). "Binaurale akustische Umgebungserkennung." In: *Fortschritte der Akustik - DAGA*. Berlin, Germany, pp. 625–626.

R. SCHARRER and J. FELS (2013). "Sound Field Classification in Small Microphone Arrays." In: *Fortschritte der Akustik - AIA/DAGA*. Meran, Italy.

R. SCHARRER and M. VORLÄNDER (2010). "Blind Reverberation Time Estimation." In: *International Congress on Acoustics, ICA*. Sydney, Australia.

R. SCHARRER and M. VORLÄNDER (2011). "The coherence estimate function and its dependency on the room acoustic situation." In: *Fortschritte der Akustik - DAGA*. Düsseldorf, Germany, pp. 625–626.

Supervised Bachelor Theses

C. HAAR (2012). "Evaluation of Binaural Reproduction Techniques with In-Ear Systems Bewertung von binauralen Wiedergabeverfahren mit In-Ear-Systemen." Bachelor Thesis. RWTH Aachen University.

K. HORNUNG (2012). "Datenbank für binaurale Aufnahmen und Raumimpulsantworten." Bachelor Thesis. RWTH Aachen University.

A. HUBER (2011). "Psychoakustische Auswertung von Langzeitmessungen." Bachelor Thesis. RWTH Aachen University.

J. TUMBRÄGEL (2013). "Implementation and Evaluation of 3D Monitoring for Musicians." Bachelor Thesis. RWTH Aachen University.

Supervised Diploma and Master Theses

H. E. SCHMOCH (2011). "Blinde Schätzung der Nachhallzeit aus der räumlichen Kohärenz." Diploma Thesis. RWTH Aachen University.

A. SILLNER (2009). "Binaural models for sound-source localization." Diploma Thesis. RWTH Aachen University.

Acknowledgments

I would like to take this opportunity to thank my family, friends, and colleagues for their support and inspiration during my time at the Institute of Technical Acoustics and especially while writing this thesis.

Thanks go to Prof. Dr. rer. nat. Michael Vorländer for giving me the opportunity to work at the Institute of Technical Acoustics and for always having an open mind for all of my ideas. I would also like to thank Prof. Dr.-Ing. Peter Vary, head of the Institute of Communication Systems and Data Processing, for taking on the role as the second examiner of my thesis.

Further thanks go to Dr.-Ing. Gottfried Behler and Prof. Dr.-Ing. Janina Fels for inspiring discussions on scientific as well as non-scientific topics and for the always-open ear for all organizational questions. I would also like to thank the staff at the electrical and mechanical workshop of the ITA, namely Rolf Kaldenbach, Hans-Jürgen Dilly, Uwe Schlömer and Thomas Schäfer, for their support and patience when dealing with different challenges and, sometimes last-minute, tasks in the context of my Ph.D. thesis.

I always appreciated the warm and friendly working environment at the ITA. This part is due to all the dear colleagues that are always helpful and friendly. I am very thankful for making the time at the ITA such a nice one. Special thanks go to the *REWE* crew, namely Josefa Oberem, Ramona Bomhardt, Martin *Joe* Guski, Marc Aretz, and Rob Opdam for the always-amusing lunch breaks. Pascal Dietrich deserves thanks for long discussions and insights on theoretic questions; Martin Pollow for deep discussions on science, philosophy and, of course, tea. Further thanks of course to all current and former ITA employees for the nice working environment. Special thanks in this context go to João Henrique Guimarães and Sebastián Fingerhuth for convincing and enabling me to join the institute in the first place.

I am also very grateful to the students that contributed to my works. Especially Axel Sillner and Hinrich Schmoch, who did do a very great job on their diploma theses, and their excellent results contributed to the final results presented in this thesis.

I also would like to thank my project partner, the company Phonak, for supporting large parts of my work; especially Alfred Stirnemann for giving me the possibility to define a project that was hopefully very fruitful for both sides. Martin Kuster gave me plenty of precious insights in his work, and always had an encouraging way in questioning my decisions and results, this sparring was quite inspiring and definitely improved the overall quality of this thesis.

Special thanks go to my family for always supporting me. Especially my wife and my son had to endure quite some mood swings due to doubts on my own work and sleep derivation. I hereby apologize for all that, and I am very grateful for your inspiration and reminding me of the really important things in life.

Curriculum Vitae

Personal Data   Roman Scharrer

5th of May 1982   Born in Meerbusch, Germany

Education

1992 – 2001   Secondary School
Albert-Einstein-Gymnasium, Kaarst, Germany

10/2002 – 10/2007   Master's degree in Electrical Engineering (Dipl.-Ing.),
RWTH Aachen University

03/2008 – 07/2013   Ph.D. studies in Electrical Engineering
RWTH Aachen University

Professional Experience

8/2000 – 12/2004   Establishment of the Company *Satisfaction Designs*
Web-design for local clubs and small businesses

3/2005 – 9/2006   Student Research Assistant
Institute of Technical Acoustics, RWTH Aachen University

10/2006 – 3/2007   Internship
Audi AG, Neckarsulm, Germany

3/2008 – 4/2013   Research Assistant
Institute of Technical Acoustics, RWTH Aachen University

Bisher erschienene Bände der Reihe

Aachener Beiträge zur Technischen Akustik

ISSN 1866-3052

Alle erschienenen Bücher können unter der angegebenen ISBN-Nummer direkt online (http://www.logos-verlag.de) oder per Fax (030 - 42 85 10 92) beim Logos Verlag Berlin bestellt werden.